D0905289

Implementing Physical Protection Systems:
A Practical Guide

David G. Patterson, CPP, PSP

ABOUT THE AUTHOR

David G. Patterson, CPP, PSP, is the founder of Patterson & Associates, an international corporate security consulting firm. He is also director of the consulting group of the Steele Foundation, an international risk management firm. Mr. Patterson has over 25 years' experience as a corporate safety and security manager and consultant for the worldwide operations of high technology companies, industrial companies, schools, and the United States government. He is a recognized author and trainer with the ASIS International Physical Security Council in the fields of security systems integration, project management, policies and procedures, safety, and business continuity planning. He is also on the faculty for the new Physical Security Professional (PSP) certification program. Mr. Patterson has conducted in-depth anti-terrorism assessments, risk analyses, and vulnerability studies, and he has developed and implemented security and safety policies, standards, and procedures for several organizations. In addition, he has designed complex access control, closed-circuit television, intrusion detection, and alarm systems for large industrial sites, seaports, offshore facilities, airports, schools, and building complexes.

Mr. Patterson is a recognized expert in safety, security, and loss prevention. He consults with business owners and managers, school administrators, security managers, and police on how to reduce risks. He has published numerous articles on security and frequently speaks at seminars and conventions on security matters. He has performed numerous crime demographic studies and vulnerability assessments of properties. In addition, Mr. Patterson is qualified as a security expert witness in litigation involving allegations of crime foreseeability, inadequate security, faulty security systems, premises liability, and security officer negligence.

Mr. Patterson's professional designations include the following: Certified Protection Professional (CPP), Physical Security Professional (PSP), Certified Fraud Examiner (CFE), and Certified in Homeland Security, Level III (CHS-III).

FOREWORD

In a lifetime, if lucky, one is blessed with a few choice individuals to call "friend." If a bit luckier, an average person can meet and shake hands with an extraordinary individual who can teach him something important. If life is truly blessed, an average man like myself can have the extraordinarily good fortune to call a man like David Patterson, CPP, a good and faithful friend and teacher.

There are leaders, followers, managers, and individuals in every aspect of our lives. Occasionally, an individual will step off to the side and share his life's experiences with the rest of us. Assuming that such testimonies are worthwhile, as are David's, it is safe then also to assume that those who read, apply, and live by the guidelines set out will benefit and prosper. Such is the case for all those who read the following pages. David Patterson's life experiences and learned lessons in security will save us time, energy, and potentially life-threatening errors.

Thank you, David, for taking the time once more to enrich my life and career.

Charlie R.A. Pierce
President
LeapFrog Training & Consulting

ACKNOWLEDGMENTS

Early in 2003, my friend and colleague, Joseph R. (Bob) Granger, asked me to help him develop the course material for the newly established Physical Security Professional (PSP) certification examination review course. We spent several months developing the material and then taught the course twice in 2003. While attending the ASIS International Seminar and Exhibits in New Orleans in September 2003, Bob arranged a meeting with the ASIS Professional Certification Board (PCB) to discuss writing a book to cover the subject matter required for the course and examination. At the meeting, the members of the PCB explained the need for additional reference material for the program and encouraged me to write a book that would specifically address Domain Three: Implementation of Physical Security Measures. I was very excited about writing the material and felt it would be an easy task since I had already developed the review course material, consisting of some 84 slides and notes. I agreed to provide a draft of the book by the end of November 2003, and my first attempt at being a book author was begun. It did not take long for me to realize how much more difficult it is to write a book about something than it is to prepare a presentation about the same subject. I soon learned that I needed a great deal of help from other authors and practitioners in the security business.

I first want to acknowledge my wife and business partner, Leslie, for her tireless efforts in encouraging me to complete the material and for reviewing the content, grammar, and format endless times—in order to meet the deadline.

I want to express my thanks and gratitude to Bob Granger and the ASIS International staff for their belief in me and for offering me the opportunity to create this book. I also want to thank Charlie Pierce for his encouragement, review of sections, and contribution of a foreword to the book. Thanks also go to Grant Angell of Limited Voltage Systems in Gladstone, Oregon, for technical guidance; to Jerry Hallett of Bayview Security, Inc., in Novato, California; and finally to Greg Pearson and Tom Stallings of the Steele Foundation in San Francisco for their review of several sections.

David G. Patterson

PREFACE

The purpose of this book is to guide security professionals in implementing physical protection systems (PPSs). The book is primarily intended as study material for the ASIS International Physical Security Professional (PSP) certification examination.

Chapter 1 defines PPSs and their three major functions: preventive, corrective, and detective. Chapter 2 discusses various threat conditions and the measures that should be introduced at each level. Chapter 3 addresses the complex subject of project management, including modern management principles that help managers accomplish projects on time and under budget. Chapter 4 provides guidance on planning issues related to implementing PPSs. Chapter 5 outlines the important precepts of designing PPSs and preparing the design specifications that will be sent to potential bidders during the procurement phase. Chapter 6 provides some techniques for estimating PPS costs, including life-cycle costs. Chapter 7 addresses the process of procurement, including the use of requests for proposals, sole source agreements, and invitations for bids. Chapter 8 examines installation issues and the development of operating procedures. The chapter stresses that these procedures should be developed during the system design process just as the hardware and software functions are. System tuning is also addressed in the chapter. Chapter 9 reviews the training required when implementing physical protection systems. Chapter 10 addresses the importance of testing PPSs, defines the different types of testing that should be performed, and examines warranty issues. Finally, Chapter 11 discusses the importance of maintenance and maintenance contracts in keeping PPSs operating effectively. It also raises concepts to consider regarding evaluation and eventual replacement of physical protection systems.

I hope you find this book helpful and instructive.

David G. Patterson
December 2003

Contents

CHAPTER 1:

PHYSICAL PROTECTION SYSTEMS

To protect a company and its assets, the very first step is to perform a threat and vulnerability analysis. Based on that analysis, the security team should implement physical protection systems (PPSs) to provide safeguards that mitigate the threats. This chapter provides an overview of the types of physical protection systems that security professionals should implement to diminish the vulnerabilities they identify.

In *The Design and Evaluation of Physical Protection Systems*, Mary Lynn Garcia states that "a physical protection system (PPS) integrates people, procedures, and equipment for the protection of assets or facilities against theft, sabotage, or other malevolent human attacks." She defines PPS functions as detection, delay, and response.

The present book divides security measures into three types:

- **Preventive measures**, which reduce the likelihood of a deliberate attack, introduce delays, reduce vulnerabilities, or otherwise cause an attack to be unsuccessful

- **Corrective measures**, which reduce the effects of an attack and restore the facility to normal operation

- **Detective measures**, which help discover attacks and activate appropriate preventive or corrective measures

Terrorist attacks and other crimes are the product of intentions and means. Technological systems help check for means but not for intentions. Technology for managing risk to vital facilities is a two-edged sword. Implemented properly, it can significantly augment capabilities. Implemented poorly, it can actually reduce the level of security. The human element of security, which relates to operating procedures, decision making, common sense, and awareness, must be included in the implementation process during the time of the system design—not after the system has been designed and implemented. Throughout its chapters, this book will emphasize the importance of properly blending technology and human factors in implementing effective physical protection systems.

Physical protection systems consist of a proper mixture of the following:

- **Architectural elements**, such as barriers and locks, exterior and interior lighting, critical building services, space layout, parking, dock facilities, egress stairs, and adjacent facilities

- **Operational elements**, such as organization and staffing, policies and procedures, training, visitor control, security guard staffing, post order assignment and execution, alarm and incident assessment, incident responses, administration of security systems, delivery processing, and emergency response

- **Security systems elements**, such as automated access control systems, intrusion detection and alarm systems, closed-circuit television (CCTV) systems, communication systems, and security control center equipment

Preventive Measures

Preventive measures protect vulnerable resources, introduce delays, and make an attack unsuccessful or reduce its impact. They include both physical and psychological deterrents. Physical security measures such as fences, barriers, locks, and window grills are physical deterrents. CCTV cameras and visible security officers trained in effective policies and procedures are psychological deterrents.

As in all health, safety, and security considerations, prevention is the most desirable option. The key to preventing security incidents is enhancing existing security programs and implementing reasonable security measures on the basis of vulnerability assessments.

Initially, an organization should analyze its facilities and operations to identify significant hazards and exposure potential and to determine the risks of a security incident. This analysis should not be limited to certain facilities or certain operating times but should include the entire scope of operations. The next step is to identify critical control points (the locations, processes, functions, or times at which the operation is at greatest risk) and establish monitoring procedures to ensure that operations are not interrupted. Since it is impossible to eliminate all hazards, the approach must employ reasonable procedures, along with documentation and verification. The steps are as follows:

- Evaluate significant threats or hazards and exposure, and estimate the likelihood of occurrence. Determine critical control points.

- Develop and implement preventive measures to reduce hazards. These preventive measures will be a combination of architectural, operational, and systems measures.

- Develop written security monitoring procedures for each critical control point. Monitoring procedures are systematic, periodic activities for ensuring that critical controls are in place and have not been breached or compromised in any way. Monitoring is normally performed by the facility guard force through patrolling the critical control points or observing those points on CCTV.

Facility

Preventive measures at a facility typically protect perimeter access, such as fencing, lighting, and vehicle barriers. Architectural measures minimize the number of exterior entrances to the building. They also direct personnel entering the building or entering restricted areas through points where entry may be controlled by either personnel or access control systems. Locks and metal or metal-clad exterior doors are examples of measures that delay attacks on a facility.

Buildings

Buildings of various types represent the most common barrier for protecting facilities. In general, building materials should be solid and offer penetration resistance to, and evidence of, unauthorized entry. Shatter-resistant, laminated glass may be used if visual access is required. Construction hardware should meet clear specifications; for example, hardware accessible from the outside should be required to be peened, brazed, or spot-welded to preclude tampering or removal.

Doors should offer resistance to forced entry. Reinforcement may also be needed for doorjambs, louvers, and baffle plates. Windows, when relied on as physical barriers, must be constructed of shatter-resistant, laminated glass of a specified minimum thickness. They must also be installed in fixed frames so that panes are not removable from the outside. It is essential that window frames be securely anchored in the walls and that windows can be locked from the inside. Under certain conditions, unattended apertures should be alarmed or equipped with steel wire mesh and steel bars with steel crossbars, which should be checked for integrity during security officer patrols.

Lighting

Interior and exterior security lighting is both a preventive and deterrent measure. Lighting makes it more difficult for an adversary to approach the facility or carry out an internal attack without being observed. This subsystem includes lighting fixtures, mounts, enclosures, and ballasts. Effective lighting provides a deterrent to adversary intrusion, helps the protective force locate and assess alarm initiations, and provides for effective use of CCTV as a surveillance and assessment tool. Lights need a minimum specified luminance at ground level for specific areas, and they require a regular power source and emergency backup power. Lights should not cause glare or bright spots in CCTV camera images, especially if CCTV is the primary means of assessing alarms.

Barriers and Locks

Physical barriers are used to control, impede, or deny access and to direct the flow of personnel and vehicles through designated portals. Barrier system effectiveness is measured against specific standards and performance testing. Specifically, barriers reduce the number of entry and exit paths, facilitate effective use of protective force personnel, delay the adversary so the threat can be assessed, protect personnel from hostile actions, and channel adversaries into preplanned neutralization zones.

Fencing. Fencing is normally used to enclose security areas and to designate property boundaries. Depending on the level of security needed, fences may require regular patrolling, continuous observation, or an intrusion-detection system supported by an alarm monitoring and assessment capability. Fences should meet specific gauge and fabric specifications, be topped with particular wire and outrigger configurations, include steel posts with bracing, and meet certain height and location provisions.

Vehicle Barriers. Vehicle barriers are used to deter penetration into areas when such access cannot otherwise be controlled. Vehicle barriers may include pop-up barriers, cables, bollards, or natural terrain obstacles (for example, bodies of water, ravines, berms, steep hills, or cliffs).

Locks. The requirements for security locks vary according to the assets being protected, the identified threat, existing barriers, and other protection measures. The General Services Administration (GSA) establishes standards for security containers. For classified matter in storage, security classification is the only factor that determines the degree of protection required. In other settings, other considerations apply, such as strategic importance, susceptibility to compromise, effect on vital production, health and safety factors, and replacement costs.

Active Denial Subsystems

Active denial systems include obscurants (smoke) and other dispensable materials, such as foams, sprays, and irritant agents. Other systems may incorporate flickering light or intense sound systems to delay, confuse, or otherwise hamper adversaries.

Employees

Employees are a company's greatest asset, but they can also be its worst enemy if the company does not know whom it hires and fails to keep a watchful eye over employees' work. The following are some preventive measures that apply to employees:

Employee Screening

- Establish policy and procedures for screening all new employees. Apply the procedures equally to all staff, regardless of race, national origin, religion, and citizenship or immigration status.

- Verify the work references, addresses, and phone numbers of all current employees, including seasonal, temporary, contract, and volunteer staff.

- Conduct periodic follow-up background checks to ensure that an employee's situation has not changed.

Employee Work Assignments and Supervision

- Map employee work assignments and make sure supervisors are aware of who should be in what locations in the building and what work assignments they should and should not perform.

- Adjust the span of control so supervisors can monitor employees closely.

- Caution supervisors to be alert for suspicious or unusual behavior.

- Provide training to promote security awareness. Teach employees how to prevent, detect, and respond to malicious, criminal, or terrorist actions.

- Provide periodic reminders of the importance of security procedures.

- Implement a balanced system of consequences and rewards and include security compliance in job performance standards.

Protective Force

A well trained protective force is an effective preventive measure. The presence of security officers may deter attacks, or officers may see attacks in progress and delay or intercept the perpetrators before they can attack the facility. The protective force is also a detective measure, as security officers can discover abnormal situations and take action to thwart attacks. Of course, the protective force also responds to security incidents and takes corrective action to restore the facility to its original state.

Visitor Controls

Visitors can represent a risk to the facility, so it is important to implement preventive controls like the following:

- Restrict visitor entry.

- Verify the identity of unknown visitors.

- Require escorts to critical and restricted areas.

- Restrict access to employee locker rooms.

- Establish systems to ensure that there is a valid reason for the visit before providing access to the facility. Beware of unsolicited visitors.

- Inspect incoming and outgoing vehicles, packages, and briefcases for suspicious, inappropriate, or unusual items or activity.

Corrective Measures

Corrective measures reduce the effect of an attack by allowing a rapid and effective response. Development of corrective measures should focus on various threat scenarios. The corrective actions should be proportional to the identified threats.

Corrective measures require both plans and personnel. The plans are of two types: (1) instructions for monitoring the PPSs, assessing the alarms and information provided by all subsystems, and dispatching an appropriate response, and (2) scenario-based procedures for responding to breaches in security. Response personnel include the protective force assigned to the facility and any local, state, or federal responders.

The security manager should develop procedures for responding to security problems or failures that occur if a critical control has been breached or compromised. These procedures should also include investigating how the breach occurred and taking action to correct any problems identified. After making corrections, security staff should revise the assessment and response procedures and retest them. They must also verify or test the procedures periodically to ensure that they work. Having confidential, written protocols is vital, and they should be revised as operations change. Chapter 8 discusses operating procedures in more detail.

The security manager must also address emergency preparedness planning and emergency response to deliberate attacks and natural or other disasters. An emergency response plan should allow swift action to assist affected personnel and to minimize the adverse consequences of the incident on personnel who are not directly affected. The following steps may help security managers prepare for crises:

- Become familiar with the emergency response system in the community.

- Develop an internal–external communications system to contact management, internal security staff, and local law enforcement personnel.

- Develop a strategy to prepare for and respond to malicious, criminal, or terrorist threats and actions.

- Plan for an emergency evacuation. Consider ways to prevent security breaches during evacuation.

- Store any response plans, floor plans, and flow plans in a secure, off-site location.

- Develop a strategy for communicating with employees, the public, and the media.

- Identify a media spokesperson, prepare generic press statements, develop background information, and coordinate press statements with appropriate authorities.

- Identify personnel who will be responsible for carrying out these measures, and determine who can serve as backup personnel.

Detective Measures

Detective measures discover attacks and activate preventive or corrective measures. Detective measures use security subsystems, written procedures, and human resources.

Security subsystems use technology to help detect the means that perpetrators use to carry out an attack on a facility. Security subsystems at facilities include display and assessment, intrusion detection, identification, centralized monitoring, access control, CCTV, communications, search, and investigation.

Display and Assessment

This subsystem consists of consoles, workstations, computers, printers, recorders, communication equipment, and displays for monitoring all the security subsystems, assessing any unusual conditions, and dispatching appropriate response force. This subsystem aids in alarm assessment, allows the protective force to track intruder progress toward a target, and aids in assessing intruder activity and characteristics (such as the number of intruders and whether they are armed). The protective force is usually responsible for monitoring alarms, responding to alarms, preparing alarm reports, and distributing copies as appropriate. Response procedures are usually found in site or facility protection plans.

Intrusion Detection

An intrusion detection subsystem alerts the internal monitoring center or a monitoring company, as well as the local police, of an intrusion attempt through a door or window at the facility. This subsystem is cost-effective compared to security guards. The main purpose of an intrusion detection subsystem is to alert the protective force to an intrusion.

Intrusion detection subsystems generally consist of both an alarm and an assessment capability. These subsystems are usually layered for both interior and exterior applications. Exterior systems are designed to provide the earliest possible detection of an unauthorized intrusion, as far away from the facility as possible. Interior intrusion detection systems may be further divided into layers according to the configuration of security areas and the required levels of protection. In addition to patrols and visual surveillance provided by the protective force, alarm and detection devices are fundamental components of any PPS.

To be effective, alarms must be clearly audible. Similarly, alarm displays must be clearly visible, must identify the location and type of alarm, and must feature an interface that allows for alarm recognition by the operator. Alarm devices require continuous supervision to preclude any covert attempt to bypass the alarm system and to ensure an appropriate and timely response. To achieve an acceptable degree of assurance that the PPS works properly, security managers should provide adequate equipment, an effective testing and maintenance program, and a sufficient number of trained personnel to operate the intrusion detection equipment.

Exterior

At some critical facilities, the outermost layer of the exterior system is a perimeter intrusion detection subsystem. It typically consists of complementary electronic sensors, such as microwave, infrared, and electric field sensors. It might also employ fence disturbance detectors and seismic sensors. Exterior subsystems must be capable of withstanding the environmental conditions in which they are deployed. Properly designed subsystems generally use two or more types of sensors, depending on the operating environment and design parameters. The purpose of the exterior sensor is to provide assurance that a person crossing the perimeter will be detected, whether walking, running, jumping, crawling, rolling, or climbing at any point in the detection zone, beyond specified weight and speed limits. Sensor subsystems need adequate coverage in all weather and light conditions. The sensor fields should overlap to eliminate dead areas, and they must be wide enough to deter bridging (that is, using a ladder or sawhorses to construct a "bridge" for walking over the detector's field). Also, it is essential that detection zones contain no dips, high ground, or obstructions that could provide a pathway for an individual to avoid detection.

Interior

Interior intrusion detection subsystems are normally designed to protect specific areas within facilities, such as computer rooms, vaults, or rooms containing high-value items. These systems employ technologies that detect physical movement, heat, speed, cable tension, vibration, pressure, and capacitance. Protective forces can assess the alarms by monitoring the systems and observing alarm areas with CCTV cameras equipped with pan, tilt, and zoom features.

Intrusion detection sensors may be either active or passive. Active sensors transmit energy into a protected area and detect changes in that energy caused by the heat or motion of an intruder. Active sensors are very effective at discriminating nuisance alarm sources. Passive sensors emit no energy of their own. They monitor the protected area and detect energy emitted by an intruder or disturbances in energy fields caused by the intruder. One advantage of passive sensors is that they are more difficult for intruders to discover than are active sensors. In addition, passive sensors are generally safer for use in explosive or hazardous environments since they emit no potentially explosion-initiating energy.

Identification

Identification systems usually employ a badge that an employee wears while in the workplace. It is important that ID badges feature employee photographs, individual control numbers, and color codes for areas of authorized access. Badges should also employ anticounterfeiting measures. Security staff should retrieve badges when employees or contractors are no longer associated with the organization.

Access Control

An access control system is both a preventive measure and a detective measure. It prevents attacks by making it more difficult for unwelcome individuals to enter the facility. Such a system also notifies security staff when unauthorized entry attempts are made.

Frequently, intrusion detection and access control are separate subsystems, interfaced to provide information to the centralized monitoring system operator. In many systems, normal access control and other work-related activities are processed without operator interaction. Records of such transactions are generally recorded for historical purposes. The key elements in establishing access control systems are as follows:

- Use single-entry devices, such as turnstiles or revolving doors, to eliminate tailgating.

- Establish automated access control systems using secure card keys that are not easily counterfeited.

- Restrict access so staff can enter only those areas necessary for their job functions and only during appropriate work hours.

- Periodically reassess levels of access for all staff.

- When an employee or contractor is no longer associated with the organization, collect his or her card key, change all card keys, or rekey locks, as applicable.

Closed-Circuit Television

A closed-circuit television (CCTV) system offers the capability to maintain visual surveillance of the site from a remote source, such as a monitoring company or a customer-operated monitoring site. In addition, a CCTV system typically records images so that security staff can reconstruct what occurred at a facility and identify the culprits involved in an incident. Cameras should be installed in parking areas, reception areas, critical processing areas, storage areas, loading areas, and various building entrances and exits. Modern subsystems use digital recording, storage, and playback devices.

Most critical facilities use fixed-position CCTV camera coverage for timely assessment of alarms generated around the perimeter of the facility. Alarms normally annunciate in the central monitoring area, where the alarm console operators can acknowledge the alarm, assess its cause, and direct a response as necessary.

CCTV systems used in conjunction with intrusion detection or access control systems are most effective when they can automatically call the operator's attention to an alarm-associated camera display with a picture quality, field of view, and image size that enable the operator to recognize a human presence. Tamper protection and loss-of-video alarm annunciation are essential if the cameras serve as the primary means of alarm assessment. Video recorders are most useful when they operate automatically and are rapid enough to record an intrusion accurately. Video capture systems provide pre-alarm, alarm, and post-alarm video images of the alarmed area to help staff determine the cause of alarms and track intruders.

Communications

The communications subsystem provides the capability for all security subsystems to work together. The communications network is often a closed private network used only by security. In other cases, the security subsystems share the same network used to support the company information technology (IT) functions. Other parts of the communications subsystem, such as intercom, radio, and telephone equipment, provide the means to report problems and request assistance when attacks are discovered.

Search Equipment

Search equipment helps detect the presence of firearms and explosives on personnel or in containers, mail, and vehicles. This subsystem includes metal detectors, explosives detectors, and X-ray machines.

Investigation

Investigative measures are the detection measures carried out by the protective force. The following are some of the main investigative steps:

- Conduct routine security checks of the premises, including utilities and critical computer data systems, for signs of tampering or malicious, criminal, or terrorist actions. Also inspect any other areas that may be vulnerable to such actions. Check restrooms, maintenance closets, personal lockers, and storage areas regularly for concealed packages or other anomalies.

- Randomly check or test critical processes.

- Investigate specific threats and any information about or signs of tampering or other malicious, criminal, or terrorist actions.

- Investigate malicious statements by employees or visitors.

- Immediately investigate missing items or other irregularities (if outside the normal, expected range), and alert appropriate law enforcement authorities about unresolved problems.

- Evaluate the lessons learned from past security incidents or other malicious, criminal, or terrorist actions and threats.

- Perform random security inspections of all appropriate areas of the facility using knowledgeable in-house or third-party staff. Verify through audits that security contractors are doing an appropriate job.

References

Garcia, Mary Lynn (2001). *The design and evaluation of physical protection systems.* (pp. 1-7). Boston: Butterworth-Heinemann.

General Services Administration. Federal Specification FF-P-2827 (PL655 and PL656) in Federal Supply GSA Schedule GS-O7F-0368J under SIN 246-36, LOCKING DEVICES. Washington. February 20, 2003.

U.S. Department of Energy (2000). *Physical security systems inspectors guide.* Washington.

Chapter 2:

Threat Conditions and Security Operating Levels

The U.S. Department of Homeland Security has developed the Homeland Security Advisory System (HSAS), which establishes various threat conditions to characterize the relative risk of terrorist attacks. Associated with each threat level are recommended protective measures that the government and the private sector should take to reduce vulnerabilities. The five threat conditions are also referred to as "operating levels" because of their hierarchical presentation. Just as the U.S. government has established threat conditions to communicate the relative risk of terrorist attack, private organizations should establish similar security operating conditions based on their perceived risk.

To assist industry, ASIS International recently released a draft document titled "Threat Advisory System Response Guidelines." The operating levels established by companies should include the criteria recommended by the government and ASIS International, but they should go further and address threats against individual facilities, such as weather-related, geographical, political, religious, or socio-logical factors unique to a location and specific business. A list of clearly defined security response activities should be adopted throughout the company to ensure that during a period of increased security concern, a considered, appropriate operational response is implemented.

Security Operating Levels

Threats against a company and its assets are always present. What fluctuates is the likelihood that one or more of the threats will be carried out. Therefore, organizations should design their facility protection plans according to a multilevel protocol. The following five-level security operating protocol was developed by the Steele Foundation and is used here by permission. This protocol is often recommended for commercial and industrial facilities:

- Levels 1 and 2: Low or Normal Business Mode (accepted level of risk; low or general terrorist risk)

- Level 3: Elevated Risk Mode (accepted level of business risk but elevated terrorist risk)

- Level 4: High Risk Mode (high risk of terrorism with specific targets identified)

- Level 5: Severe Risk Mode (confirmed threats with specific locations and time frames)

Level 1 and II: Low or Normal Business Mode

These are the modes in which companies should establish their normal security program, based on the following factors:

- Prominence of the organization or the buildings it uses

- Past crimes committed by employees of the firm

- Crime profile of the areas where buildings are located

- Threat and consequences of international and domestic terrorism

- Threat and consequences of workplace violence

- Liability, legal issues, and increased insurance premiums arising from safety and security incidents

- Public relations and credibility issues

- Costs of dealing with a crisis

- Lost revenue from business interruption

- Loss of confidence of clients and shareholders

- Impact on employee morale and productivity

Level 3: Elevated Risk Mode

When implementing security measures appropriate to this mode, organizations should consider the following:

- Rumors of terrorist or other threats received by the company

- General terrorism or other threats validated by a law enforcement or government agency

- Major terrorist-related or other incidents in the local area

- Political or civil unrest in and around the facility

- Threats against other company facilities

- Increased concerns of natural disaster

Level 3: Heightened Risk Mode

When implementing additional measures in response to this highest level, companies should consider the following:

- Threats against the company or specific tenants in the company's current office locations

- Terrorist or other threats against adjacent properties

- Major terrorist or disruptive activity carried out in the vicinity

- Confirmed threats of natural disaster

Level 4: High Risk Mode

To implement additional security measures, companies should consider the following:

- Targeted threats against the company or specific tenants in their current office location

- Terrorist or other threats against adjacent properties

Level 5: Severe Risk Mode

As they implement additional measures, companies should consider these factors:

- Major terrorist or other disruptive activity already carried out in the local vicinity

- Confirmed threats of natural disaster or other threat activity

Facility Protection Plan

Once the organization has established operating levels for each facility, the next task is to develop facility-specific protection plans. A facility protection plan is a narrative document that describes the perceived threats, their impact on the facility and on the organization, all the security countermeasures to be applied. The plan starts at the perimeter of the property and then moves inward to cover the doors, windows, roof access, utility access, and other facility elements. A different plan is developed for each operational level. Tables 2-1, 2-2, 2-3, and 2-4 summarize sample measures employed at each operational level.

Table 2-1
Levels 1 and 2: Low or Normal Business Mode

Security Measure	Business Hours	After Hours
Perimeter	Main lobby guard monitors traffic. No vehicles allowed to park close to the building.	Security desk guard monitors traffic. No vehicles permitted to park close to building.
Main Lobby Doors	Open.	Locked and card access only. Employees expected to use their access cards to exit.
Other Perimeter Doors	Card access enabled. Free egress.	Card access enabled. Free egress.
Access Control Turnstiles	None.	None.
Lobby Operational Controls	Security guards monitor personnel entering and direct visitors.	Security guards monitor personnel entering. No visitors allowed.
Elevator Controls	Service elevator on access control.	Service elevator on access control.
Visitor Controls	Visitors must sign in to receive sticker badge and must be escorted.	Visitors entering through security door are required to sign visitor log.
Parking Garage Controls	Access control for employee parking. Intercom and screening procedures for visitors and employees who have forgotten access cards. Reserved area for visitor parking. No control over visitor parking. Security officers patrol garage and cite vehicles parked in reserved spots.	Access control for employee parking. Intercom and screening procedures for visitors and employees who have forgotten access cards. No patrol.
Loading Dock	Access control for loading dock doors. Vehicles screened at bottom of ramp.	Loading dock closed. Overhead door closed and protected by access control.
Mail and Package Handling	All mail and packages delivered through loading dock. Delivery personnel must present ID and sign log.	Deliveries received at security desk.

Table 2-2
Level 3: Elevated Risk Mode

Security Measure	Business Hours	After Hours
Perimeter	Main lobby guard monitors traffic. No vehicles allowed to park close to the building.	Security desk guard monitors traffic. No vehicles permitted to park close to building.
Main Lobby Doors	Open.	Locked and card access only. Employees expected to badge out at security desk.
Other Perimeter Doors	Card access enabled. Free egress.	Card access enabled. Free egress.
Access Control Turnstiles	None.	None.
Lobby Operational Controls	Security guards monitor personnel entering and direct visitors.	Security guards monitor personnel entering. No visitors allowed.
Elevator Controls	Service elevator on access control.	Service elevator on access control.
Visitor Controls	Visitors must sign in to receive visitor badge and must be escorted.	Visitors entering through security door are required to sign visitor log.
Parking Garage Controls	Access control for employee parking. Intercom and screening procedures for visitors and employees who have forgotten access cards. Reserved area for visitor parking. No control over visitor parking. Security officers patrol garage and cite vehicles parked in reserved spots.	Access control for employee parking. Intercom and screening procedures for visitors and employees who have forgotten access cards. No garage patrols.
Loading Dock	Access control for loading dock doors. Vehicles screened at bottom of ramp.	Loading dock closed. Overhead door closed and protected by access control.
Mail and Package Handling	All mail and packages delivered through loading dock. Delivery personnel must present ID and sign log.	Deliveries received at security desk.

Table 2-3
Level 4: High Risk Mode

Security Measure	Business Hours	After Hours
Perimeter	Front entrance barricaded; no vehicles allowed close to building. Delivery vehicles and autos screened and searched on access roads. Additional security guard needed to accomplish screening.	Barriers deployed at main entrance and access roads; no vehicles allowed.
Main Lobby Doors	Open. Security officers screen and search personnel entering and send visitors to reception desk. May require additional officers.	Locked and on card access. Employees expected to badge out at security desk.
Other Perimeter Doors	Revolving doors installed, controlled by access cards. Egress enabled.	Revolving doors installed, controlled by access cards. Egress enabled. Intrusion alarms activated.
Access Control Turnstiles	Optical turnstiles installed in areas leading to elevators and other inner areas.	Optical turnstiles installed in areas leading to elevators and other inner areas.
Lobby Operational Controls	More security guards monitoring turnstiles, monitoring other doors, and issuing visitor passes.	Security guards monitor personnel entering. No visitors allowed.
Elevator Controls	Passenger and service elevator controls active.	Passenger and service elevator controls active.
Visitor Controls	Only preregistered visitors issued day passes. Unannounced visitors allowed only under special circumstances. All visitors logged and issued access cards. Additional officer posted in lobby to handle visitor logging and badging.	No visitors allowed unless prearranged. All visitors logged and issued access cards.
Parking Garage Controls	Access control for employee parking. Intercom and screening procedures for visitors and employees who have forgotten access cards. Reserved area for visitor parking. No control over visitor parking. Security guard patrols garage and cites vehicles	Garage closed to entry. Emergency and essential personnel vehicles allowed. All vehicles searched and personnel screened.

Security Measure	Business Hours	After Hours
Parking Garage Controls, continued	parked in reserved spots. Additional security officers placed at each garage entry point to search vehicles and request identification. All visitor vehicles searched before being allowed onto parking lots. Additional officers added for searches. Garage patrols increased and security officers added.	
Loading Dock	Overhead door closed. Barriers deployed. Dock officer verifies manifest prior to entry. Trucks are queued at the street and visually inspected by additional officer.	No deliveries accepted. Loading dock locked. Overhead door closed. Barriers deployed. Additional security officer posted in dock guardhouse.
Mail and Package Handling	Activate off-site contract delivery service to accept and screen packages. Mail and parcels screened and x-rayed. Delivery personnel screened and vehicles searched.	No deliveries accepted except in emergencies and when prearranged.

Table 2-4
Level 5: Severe Risk Mode

Security Measure	Business Hours	After Hours
Perimeter	Barriers deployed. No vehicles allowed except those of essential personnel. All vehicles searched and personnel screened.	Barriers deployed and no vehicles allowed except those of essential personnel. All vehicles searched and personnel screened.
Lobby Doors	Locked with card access. Restricted to essential personnel.	Locked with card access. Restricted to essential personnel.
Other Perimeter Doors	Locked. Card ingress disabled. Free egress. Intrusion alarm activated.	Locked. Card ingress disabled. Free egress. Intrusion alarm activated.
Optical Turnstiles	Disabled. Perimeter doors locked.	Disabled. Perimeter doors locked.
Lobby Operational Controls	If allowed entry in emergency situation, personnel are checked in and out of building at lobby security desk.	If allowed entry in emergency situation, personnel are checked in and out of building at lobby security desk.
Elevator Controls	Passenger elevator controls disabled. Service and freight elevators disabled.	Passenger elevator controls disabled. Service and freight elevators disabled.
Visitor Controls	Visitors not allowed.	Visitors not allowed.
Parking Garage Controls	Garage closed to all except essential vehicles. Barriers deployed at all entrances except emergency entrances and exits.	Garage closed to all except essential vehicles. Barriers deployed at all entrances except emergency entrances and exits.
Loading Dock	Loading docks closed. Overhead door closed. Barriers deployed.	No parking allowed. Loading dock closed. Overhead door closed. Barriers deployed.
Mail and Package Handling	No deliveries accepted except in emergencies.	No deliveries accepted except in emergencies.

References

ASIS International (2003). Threat advisory system response guidelines. Draft. Alexandria, VA.

CHAPTER 3:

PROJECT MANAGEMENT

Project management has evolved for the purpose of planning, coordinating, and controlling the complex and diverse activities of modern security projects. Several forces have driven the development of project management techniques: the exponential expansion of human knowledge, the growing demand for a broad range of complex and customized security systems, and the evolution of worldwide competitive markets for security products and services. These three forces combine to mandate the use of project teams to solve problems that used to be solved by individuals.

What Is a Project?

In his book *Winning in Business with Enterprise Project Management*, Paul Dinsmore defines a project as "a temporary endeavor undertaken to accomplish a unique process." Projects normally involve several people performing interrelated activities, and the main sponsor of the project is often interested in the effective use of resources to complete the project in an efficient and timely manner.

The following attributes of a project provide more insight into how projects differ from regular work:

- A project has a unique purpose and an explicit goal to be completed within time and budget specifications.

- A project is temporary. It has a definite beginning and end. It occurs only once, and then it is finished.

- A project requires resources, such as money, people, equipment, and supplies. They may originate inside or outside the company, cross departments, require different knowledge and skills, and involve outside contractors and consultants. Resources are limited, must fit in the budget, and must be used effectively.

- A project should have a primary sponsor or customer. That party provides the project's direction and funding.

- Every project involves uncertainty. As every project is unique, it may be difficult to define the project objectives clearly and to estimate how long it will take to complete or how much it will cost. Uncertainty is the main reason project management is so challenging, especially in security projects.

What Is Project Management?

Dinsmore's definition of project management is "the mixture of people, systems, and techniques required to carry the project to successful completion." To execute a project successfully, the project manager must balance the often competing factors of scope, time and cost.

A good project manager is the key to a project's success. It has been said, "There are no good project managers—only lucky ones." To manage a project, a person should be well organized, have great follow-up skills, be process oriented, be able to multitask, have a logical thought process, be able to determine root causes, have good analytical ability, be a good estimator and budget manager, and have good self-discipline.

In addition to good process skills, a project manager must have good people-management abilities, including the following:

- **General management skills.** These help in establishing processes and making sure people follow them.

- **Leadership skills.** These help in getting the team to follow the project manager's direction willingly. Leadership is about communicating a vision and getting the team to accept it and strive to get there with the project manager.

- **Supervisory skills.** These involve setting reasonable, challenging, and clear expectations of people and then holding them accountable for meeting the expectations.

- **Team-building skills.** If the project manager can build a team, personnel will work together well and feel motivated to work hard for the sake of the project and their other team members.

- **Oral and written communication skills.** The project manager must possess active listening skills and give performance feedback to team members.

Project managers work with stakeholders, who are the people involved in or affected by project activities, such as the project sponsor, project team, support staff, customers, users, suppliers, installers, and even opponents of the project. Successful projects have clear plans as to what they will achieve, and they provide a noticeable benefit to the company for the costs incurred. Common elements of successful projects include the following:

- Consistent involvement of end users

- Executive management support

- Clear statement of objectives and requirements

Benefits of Quality Project Management Process

Project methodology can deliver many benefits—some evident, some hidden. These are among the most important:

- Putting project goals and expectations on paper up front encourages open communication and provides a commitment and focal point for the project team throughout the process.

- The customers usually get what they ask for—a fully documented system that performs to specifications.

- Techniques leveraged from experience help to accelerate projects, reduce risk, and assure higher quality output.

- Productivity is increased by making maximum use of all resources. Each project team member is held accountable for his or her assigned tasks.

- Comprehensive training and documentation are conducted.

- Project controls help provide early warnings of time and cost overruns.

A standardized approach provides the pathway to success. Repeatable and documented work practices act as a consistent guideline for completing the project.

Planning

"Fail to plan, plan to fail." This old adage is always true in a security project. In planning the project, one documents the answers to several key questions:

- What is the real need or purpose of this project?

- What methods, processes, or actions were used to define the project?

- What will the results of this project do for the stakeholders?

- What is the priority of this project, how was that priority determined, and how does it compare with that of other current projects?

Answers to these and other questions are needed to develop accurate project objective statements and agreed-upon project results. The documentation prepared during the planning stage is the "project scope document," which should have the following sections:

- **Executive Summary.** Include a summary that will be easier for senior managers to digest than the full, lengthy project scope document.

- **Project Benefits.** Describe the business benefits of the project.

- **Project Objectives.** State the objectives that the project will achieve. The objectives should support the business goals and objectives. The deliverables produced should also help achieve the objectives.

- **Project Scope.** Add information as to what the project will and will not produce—in other words, what is in and out of scope. This will make it much easier to manage scope change during the execution stage of the project. The section on scope should list deliverables and answer these questions:

  — What business processes are in scope and out of scope?

  — What transactions are in scope and out of scope?

  — What data will the project work with, and what data is out of scope?

  — Which organizations will be affected, and which will not?

  — Which other projects are affected, and which will be left alone?

- **Estimated Project Hours.** Estimate the total staff hours required. Provide information on how the estimate was prepared and what assumptions were made.

- **Estimated Cost.** Estimate the cost for labor, based on the effort hours. Add any non-labor expenses, such as hardware, software, training, or travel.

- **Estimated Duration.** Estimate how long the project will take in days to complete, once it starts. If the start date is known, then the end date can be determined here as well.

- **Assumptions.** What external events must occur for the project to be successful? Assumptions can be identified through the experience of knowing what activities or events are likely to happen in the organization and through brainstorming sessions with customers, stakeholders, and team members.

- **Major Risks.** For each identified risk, include a specific plan to ensure that the risk does not occur.

- **Objectives.** Objectives are concrete statements describing what the project is trying to achieve. The objectives should be written so they can be evaluated at the conclusion of a project to see whether they were achieved or not. A well-worded objective is SMART—that is, specific, measurable, aggressive but achievable, realistic, and time bounded. An objective statement should state, for example, "Upgrade the access control system by December 31 to control access through all perimeter doors with an average response time of no more than two seconds."

Project Stages

It is a good practice to divide projects into several stages, as follows:

- **Stage 1: Project Feasibility.** The project team develops a description of the project. The team conducts the threat assessment and vulnerability analysis; states the project's business objectives; identifies the requirements of the PPS and various alternative approaches; makes preliminary cost estimates and justifications; and writes an overview of the work to be accomplished.

- **Stage 2: Project Development.** The project team prepares its approach to the project, defines the deliverables, develops a list of activities that have to be accomplished to produce the deliverables, and sequences the tasks to form the work breakdown structure (WBS). The team breaks the project into chunks of six months or less and ensures that each chunk is "SMART," as described earlier. The team also ties all contract payments to specific life cycle phases.

- **Stage 3: Project Execution.** The project team accomplishes the required work and provides the deliverables according to the activities list and schedule developed in Stage 2. Also completed are all parallel activities involving all parts of the organization, such as training, testing, and documentation.

- **Stage 4: Project Closeout.** The deliverables are formally accepted, and the project is closed out. The team prepares a lessons-learned report about the project experience.

Project Deliverables

Implementing physical protection systems follows a life cycle that includes several phases: planning; design; estimation; procurement; installation and operation; training; testing and warranty; and maintenance, evaluation, and replacement. Each phase in that cycle should result in certain deliverables—that is, the products of the defined work activities. Some of the project deliverables for the various phases are as follows:

- **Planning:**

Threat assessment vulnerability analysis	Business objectives
Recommended safeguards	Design criteria
PPS requirements document	Design requirements
Procurement method	Performance requirements
Sole source justification	Capacity requirements

- **Design:**

Contract information	Equipment lists
Bidders' instructions	Security devices schedules
System specifications	Door hardware
Evaluation criteria	CCTV camera schedule
Implementation schedule	Drawings

- **Estimation:**

  Budgetary estimates   Life cycle cost estimates
  Preliminary design estimates   Schedules and time frames
  Final design estimates

- **Procurement:**

  Bidders' conference   Interview results
  Technical evaluations   Due diligence results
  Cost evaluations   Contract

- **Installation and Operation:**

  Factory test plan   As-built drawings
  Acceptance test plan   Operating procedures
  Training syllabus   Response procedures
  Training manuals   Product data
  Troubleshooting guides   Commissioning plan
  Maintenance procedures   Acceptance test results
  Factory acceptance tests   Punch list
  Site acceptance tests   Project completion certificate

- **Training:**

  Training manuals and audiovisuals   Training class evaluations
  Lesson plans, agendas, and sched-
      ules

- **Testing and Warranty Issues:**

  Pre-delivery or factory accep-   Warranty plan
      tance test data   Warranty reports
  Site acceptance test data   Warranty records
  Reliability or availability test data   Upgrades
  After-acceptance test data

- **Maintenance, Evaluation, and Replacement:**

  Maintenance records   Upgrades
  Trouble reports   Operating costs and maintenance
  Review of operations logs and   records
      records   Replacement study

Activities

A list of activities or work to be accomplished is derived from the PPS life cycle stages and deliverables. Activities define the project work to be done, depict relationships between other activities, and define the work necessary to produce the deliverables. As a general rule, most activities should be designed to take one to four weeks to execute. After defining the activities for each deliverable, the project team should sequence the tasks according to how the work must be accomplished with the available resources.

Work Breakdown Structure

The work breakdown structure (WBS) should be sequenced by life cycle phases and deliverables. It determines the activities to be executed for the entire project to be completed. The point of the WBS is to capture all the elements of work. After completing the initial breakdown of the work, the project team should determine whether any of the deliverables require more effort. The team should describe the detailed steps that must be done to complete each of the deliverables and should continue to break down each deliverable until all the work is represented completely.

Network Diagram

When the WBS is complete, the team should convert it into a network diagram. This is done by taking all the detailed activities and sequencing them in a rough chronological order. This step clarifies what activities need to be done first, second, third, etc.

Once a rough sequence has been established, the project team should go through the activities again, looking for relationships and dependencies between the activities—especially whether one activity must be completed before another can start. Most activities will have some type of "finish to start" relationship. In many cases, two or more activities may need to be completed before another one can start. The network diagram shows which activities depend on other activities and what work can be done in parallel with other work.

Contingency Activities

After building the WBS and network diagram, the team should detail the project's scope and assumptions and address such issues as the project manager's authority, assumptions about the company, and support from the organization. The team should look at potential obstacles, rank them according to probability, and build in additional activities to mitigate those potential risks.

The team should take any risk events that have a schedule or cost impact and create a contingency activity named for each. The team should then examine the WBS, determine which activities would be affected by each risk event should it occur, and calculate the duration of the contingency activity. As an example, if the risk event had an estimated impact of four weeks and an estimated probability of 25 percent, then the scheduled duration for the contingency activity would be one week.

Once the dependency network is defined, the team should create a "finish to start" dependency link from each affected activity to the appropriate contingency activity. The normal successor to the affected activity should be defined as a successor to the contingency activity. This process compensates for scheduling uncertainty.

To compensate for cost uncertainty, the team should assign to each contingency activity the type of resource that reflects the cost of the impact. For example, if the impact of the risk event is $100,000 and the likelihood is 40 percent, then the team would assign a budget resource to the contingency activity and plan for a value of $40,000.

The following are some sources of risk to consider:

- Construction requirements

- Performance expectations

- Technology challenges

- Changes in priorities

- Underestimating

- Natural and man-made disasters

Parallel Activities

In developing activities, it is a good practice to note which can be done in parallel and which are dependent on other activities. These dependencies help determine the critical paths where effort must be concentrated. The most common activity sequence is "finish to start" (meaning one task cannot start until another one finishes), but schedule charts usually show "finish to finish" milestones (meaning one task cannot finish until another task finishes). Before preparing the final activities and sequence, it is important to review the resources (such as people and special equipment) that will be needed at various locations and times to avoid conflicts and ensure availability.

Some parallel activities include the following:

- Personnel training

- Access control system programming

- CCTV system interaction programming

- ID badge production (including photographing of employees)

- System assessment procedure development

- Incident response procedure development

Project Management Software

If the team has not entered the activities into a project management software tool, it should do so at this point. The larger the project, the more critical it is to use an automated tool to help build and maintain the work plan. Although the activities can be entered into the software in any order, the process makes more sense if the activities are entered chronologically. As the team enters the activities, it can also enter the dependencies, since the prior activities should already have been entered. For each activity entered, the team should also include the estimated work effort in time increments (usually days).

This is also the time to enter any date constraints. Date constraints are events that are outside the control of the project team and must be managed around. For instance, a deliverable may need to be completed before the board of directors meeting on a certain date.

Fortunately, most project management software packages calculate the critical path automatically from the network diagram. This is the sequence of activities that takes the longest time to complete. If any activity is delayed, the project completion date will be delayed.

Many project management software tools are available. Microsoft Project is one easy, inexpensive package. Figure 3-1 shows sample output from that program.

Figure 3-1
Output from Microsoft Project

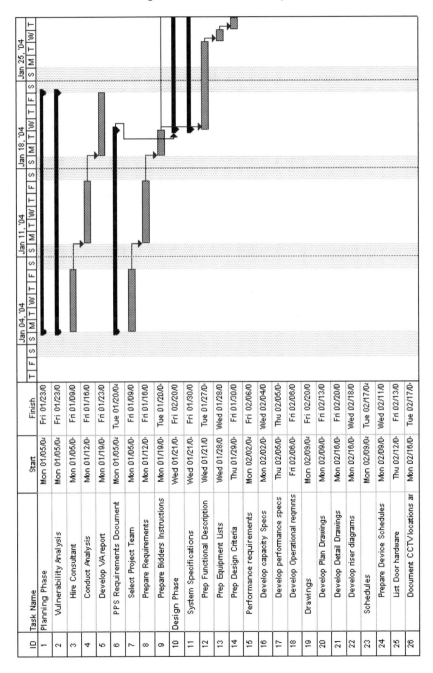

Project Team

The project team that this book has been referring to should contain a representative from each functional area. It should also include representatives from the company's information technology department. It is critical when selecting a project team that these fundamentals are represented on the team. Each team is different. In some cases one individual is the go-to person for a number of skill requirements. In other teams a certain skill is so important that the team requires many individuals to possess it. If the skills are not available, the team should be supplemented with outside consultants who possess the skills. (For example, many project teams are overloaded with technical skills and lack communication skills, which could be provided by consultants.) However, bringing in too many outsiders can prevent a team from coalescing well.

In any significant project, the team will evolve and grow. Effective project leaders have mechanisms in place to allow new team members to be successfully integrated into the team. A deep bond develops quickly within a project team. Such bonding is a powerful tool, but it can also hinder acceptance of sorely needed resources at critical points in the project. In practice, a management style that emphasizes participation and consensus is the most effective.

Project Control

Over the length of a project, what consumes the greatest amount of project management time, day in and day out, is the control effort. The effectiveness of the planning and scheduling steps is tied directly to the ability to control a project. The project manager must spend a significant amount of time regularly gathering information on what work has or has not been completed. This information must then be incorporated into the original plan to determine the differences between planned status and actual status. A key issue is to identify variances and record them for future reference and estimating purposes. With project management software, the project manager inputs the actual time and completion dates for activities, and the program automatically updates the information and identifies areas that need attention.

The project manager should require weekly status meetings and reporting of hours worked, percentage of activities completed, and estimates of the hours of work remaining to complete the assignment. This combination allows the project manager to maintain tight control while placing the responsibility for achievements on the team members. All issues discussed and decisions made at the weekly meetings should be documented. Whatever goes undocumented may as well have gone unsaid.

Important steps in controlling the project are to conduct life cycle phase and deliverable reviews each week, provide frequent project reports to stakeholders and management, use a formal change control process that includes stakeholder approvals of all changes, and revise activities and work breakdown structure to reflect changes. Positive control reveals problems early so that project issues can be better identified, tracked, managed and resolved. The project manager must also proactively manage the project scope to ensure that only what was agreed to is delivered, unless changes are approved through the change control process.

What gets measured is more likely to get done, so it is also desirable to define and collect measures of success to give a sense of how the project is progressing and whether the deliverables produced are acceptable. Finally, the project manager must manage the overall work plan to ensure work is assigned and completed on time and within budget.

Change Control Procedure Overview

During Stage 1 (project feasibility) and Stage 2 (project development), changes to the project can be made freely. When Stage 2 is complete, a project baseline is created, establishing the approved project scope, deliverables, and activities. Once the project moves into Stage 3, any changes to the project scope, deliverables, or activities must be subjected to a more systematic change process and submitted to a change review board (CRB). The CRB typically consists of representatives from each of the project's major concerned parties, such as the project manager and representatives of human resources, information technology, quality assurance, and the end users. On small projects, the CRB might consist of only one, two, or three members. On projects involving multiple departments and companies, it can swell to 30 or more members. The CRB's most important function is to serve as a central clearinghouse for changes to ensure that all important viewpoints are considered.

Any changes to the project scope, deliverables, or activities are treated systematically. Changes are proposed via change requests. A change request is a formal document that describes the project, scope, deliverable, or activity in question, the proposed change, and the impact of the change (cost, schedule, and benefit) from the point of view of the party proposing the change. Creating a change request form is a good idea on even the smallest projects because it provides a record of the project's decisions that is far more reliable than people's memories. The important elements to include in the change request form are these:

- **Change Request Number:** an arbitrary numbering scheme, usually 1, 2, 3....

- **Date Requested:** the date of the change request form

- **Requested By:** name and contact information of requestor

- **Scope Change:** brief description of the request

- **Priority:** a rating of the relative importance or priority of the request, using a high-medium-low or 1-2-3 scale

- **Justification for Change:** description of why the change is necessary and how it will benefit the project or the organization

- **Impact If Not Approved:** description of the impact on the project if the change request is not implemented, including such issues as safety, security, operability, technical performance, costs, and schedule

- **Assigned To:** statement of who is assigned to investigate the change and determine the impact to the project

- **Date Resolved:** the date on which the request was granted or denied

- **Status:** usually pending, in progress, complete, or disapproved

- **Resolution/Comments:** brief description of how the scope change was resolved

The CRB evaluates the requests to determine whether the changes are feasible in terms of time, resources, costs, and other constraints; what will be the impact of the changes on other projects; and whether to implement the changes in the current project or possibly in a future project.

The CRB identifies parties that might be affected by the change and distributes the change request for their review. The concerned parties assess the costs and benefits of the proposed change from their individual viewpoints. CRB members combine the assessments and prioritize the change request, accepting, rejecting, or deferring it. The CRB notifies all concerned parties about how the change request was resolved.

PPS Project Rules of Thumb

Some rules of thumb have been developed over the years regarding costs for physical protection systems. For example:

- Design and engineering: 10-15 percent of total cost

- Hardware and software: 15-20 percent of total cost

- Installation: 40-50 percent of total cost

- Training: 3-5 percent of total cost

- Contingency: 10 percent of total cost

Pitfalls

Even a most carefully managed project may encounter problems in execution. For example:

Management loses interest and starts borrowing project resources. In this case, the project team should do the following:

- Revalidate sponsorship.

- Better communicate accomplishments.

Little changes start adding up to a big impact (scope creep). Most project managers know they should invoke scope change management procedures if they are asked to add a major new function or a major new deliverable to a project. However, sometimes it does not seem worthwhile to invoke these procedures if the change requested is small. Over time, a series of small changes accumulates to have a significant impact on the project. If the project experiences scope creep, the project team should do the following:

- Balance constraints (in cost and time).

- Revisit the scope definition.

- Develop change requests.

The project starts missing deadlines for the completion of major activities. In that case, the project manager should do the following:

- Revisit goals and estimates with the team.

- Examine resources.

The project team loses enthusiasm. In that case, the project manager should do the following:

- Reinforce goals and management commitment.

- Revisit team selection.

- Look at leadership style.

- Reinforce coordination between activities.

Project Completion

The project is not complete until there is proof that all objectives have been achieved, along with the necessary transfer of knowledge from the project team to the operations personnel who will have responsibility for the systems after implementation. Completion requires that the project be closed down, all project knowledge transferred to the operating personnel, and all project team members returned to their home organizations. It is hazardous to let the project catch the 99.9

percent completion syndrome. The project manager should do whatever it takes to close the project out and disband the team.

PPS Implementation

The remainder of this book describes the multistep process of implementing a physical protection system. Due to the complexity of many electronic security measures, security managers must conduct a lengthy, complex process to implement them. The typical implementation process can take 18 to 24 months. If the security project is part of a larger construction project, like the construction of an office building, the project may take twice as long. Therefore, security professionals should start the process far enough in advance to avoid a crisis with the existing systems, such as escalating maintenance costs, unreliable equipment, or hopelessly outdated software.

PPS implementation has a life cycle that includes planning; design; estimation; procurement; installation and operation; training; testing and warranty issues; and maintenance, evaluation, and replacement.

- **Planning.** The planning phase produces two very important products: requirements or objectives of the newly defined physical security measures, and an operational and economic justification for the new security measures.

- **Design.** In the design phase, security staff develop all the necessary documentation to support the procurement of the physical security system. The procurement documents or procurement "package" consists of these main parts from the design phase: contract information and bidders' instructions, system specifications, and drawings.

- **Estimation.** In this phase, security staff make budgetary, preliminary design, final design, and life cycle cost estimates.

- **Procurement.** The procurement phase consists of the activities surrounding advertising for and selecting a vendor to supply and install the system. Large organizations use three common types of procurement actions: sole source, request for proposals, and invitation for bids. Each approach has its own set of rules and is useful for certain situations.

- **Installation and Operation.** This phase involves preparing the site for installation of the new system, installing the system, testing it to make it ready for use, training all personnel involved with the operation, updating and maintaining the new system, and formally accepting the system.

- **Training.** All the technological and procedural precautions in the world will be ineffective if they are not executed properly. Proper execution requires training.

- **Testing and Warranty Issues.** The tests performed by the implementation team may involve equipment, personnel, procedures, or any combination. Tests should simulate realistic conditions and provide conclusive evidence about the effectiveness of the security system. While a system is still under warranty, usually for one year, the installation contractor is expected to maintain the system free of charge.

- **Maintenance, Evaluation, and Replacement.** Once the vendor has completed the warranty commitment, maintenance of the PPS must be assumed by the customer organization. Sometimes an organization will train its own personnel to maintain the PPS. Other times, the organization will have the supplying vendor continue to perform maintenance. Another option is to solicit bids from other contractors for the maintenance work.

  At some point, the system will complete its useful life and the process of replacement will begin. To justify the replacement cost, factors such as the cost of maintenance, lack of spare parts, obsoleteness of hardware and software, operating costs, unreliability, and other negative aspects should be considered. Replacement may also be justified by new technologies and features that provide improved security, the ability to reduce manpower, or other benefits.

References

Dinsmore, Paul C. (1999). *Winning in business with enterprise project management.* New York: AMACOM.

www.allpm.com. Web site with links to many project management resources.

CHAPTER 4:

PLANNING

Project Team

The first step in the planning phase of the PPS life cycle is to form a team to develop the requirements of the physical security system. In accomplishing the planning, it is important to obtain early involvement of all stakeholders. When companies decide to procure a new information technology system, they typically form a committee of pertinent stakeholders in the company. They should do the same for a security system, including representation from all affected parts of the organization: procurement, human resources, facilities management, information technology, and critical functional managers within the organization. Every part of the organization brings a different perspective to the project. Early involvement usually means faster acceptance and more thorough implementation.

Requirements Document

One of the major outputs of this phase—the requirements document—identifies the primary reasons for implementing new measures or upgrading an old system. The reality is that the perfect system, addressing all the needs of all the stakeholders, does not exist. In evaluating and selecting a system, compromises must be made. The drafting of the requirements document provides an opportunity to identify the critical success factors of a new system. Determining the requirements that cannot be sacrificed is a means of identifying priorities, which should be stated clearly and early in the document. If the project does not have a clear focus, vendors will not understand what the organization is trying to accomplish, and the team will receive proposals that fail to address the organization's needs.

The project team should establish a calendar of activities and adhere to it. Early on, the project manager should determine the availability of key staff, taking into account vacations, holidays, and scheduled events. Vendors should be booked in advance so they can be sure to have their key personnel available to make demonstrations. The calendar will help order events, particularly those that depend on one another. At the same time, it is good not to schedule too tightly and to be realistic about the ability of the evaluation team to review proposals and to keep systems straight in their minds, especially if they must attend demonstrations scheduled closely together.

It is wise to establish a regular meeting schedule, such as every Tuesday. At a minimum, a calendar of meeting dates should be distributed to all parties involved so they can commit to meetings and decisions can be made in a timely manner.

Stakeholders

Next it is time to select the stakeholders who will help evaluate the proposals submitted by vendors. All evaluators must participate in all aspects of the selection—developing the requirements, attending demonstrations, reviewing proposals, etc. Partial participation can delay and invalidate the results.

Security systems are not purchased every year, so the project team may need to develop a clear understanding of the contracting requirements. The organization's procurement and legal departments should review and, if necessary, modify the contract language. A contract used for purchasing supplies will not work for licensing security software. The team should also find out whether the contract must be signed within a certain period to ensure full funding. There may also be legal requirements for the handling of documents. Outlining the approval process, including the times when signatures are needed, may save many weeks in later contract negotiations.

An integral part of the planning process is the collection, review, and analysis of data relative to the facility where the system will be implemented. The facility's mission, operations, and processes should be defined along with specific assets requiring protection and the physical protection measures desired. During this phase, the team should also collect information on requirements, objectives, constraints, and concepts for countermeasures.

The following are some of the types of information that should be gathered in the planning process:

Organization charts
Maps showing work areas, buildings, security posts, vital equipment areas, and material storage areas
Facility protection plans
Security plans for new areas under construction and temporary material storage areas
Plans for decommissioning existing systems
Past security survey reports
Facility asset lists
Priority of facility-specific threats

Identification of preferred methods of attack on the facility
Identification of security measures (prevention, detection, correction) that are most critical in providing protection for assets
Identification of the last possible point at which an adversary must be detected to allow adequate response time by the facility protective force
Defense-in-depth measures
Comparison of vulnerabilities against findings of past survey

To gain valuable background information, the team should visit other companies that have recently procured systems. It may be helpful to interview the staff about what worked, what did not, and what they would do differently if they could do it again.

Problem Definition

In developing a statement of the problems that are to be resolved, the project team should consider such issues as the following:

Terrorism, vandalism, and theft history in the area or for the company

Industrial or commercial break-in history in the area

The most likely target for an intruder: physical assets, information, etc.

Threats to employees from workplace or domestic violence

Employee health and safety issues

Past fraud and embezzlement problems

Protective force levels and costs

Protective force efficiency problems

Problems identifying persons entering and leaving the facility

Problems assessing and responding to alarms

Problems with surveillance of unattended sites

Problems with loss of proprietary information

Alternatives

After naming the problems to be resolved, the next consideration is to achieve a solution. Options may include any combination of the following:

Lighting

Barriers, fencing, and gates around the perimeter

Perimeter detection systems

Building intrusion detection and alarm systems

Locks and doors

CCTV surveillance and recording systems

Access control systems

Protective service posts and staffing

Centralized monitoring and control

Policies, procedures, and training

The list of alternatives depends on the circumstances and requirements at the particular site, but it is important at least to make a list, consider all the possibilities, list pros and cons for each possible solution, and develop cost estimates.

The next step is to develop the objectives for implementing or upgrading the physical protection system. The objectives must be stated clearly and be SMART (specific, measurable, aggressive but achievable, realistic, and time-bounded). Examples of such objectives include these:

- The PPS will reduce crime and vandalism at the site by 10 percent through controlling access and monitoring intrusions.

- The PPS will detect intruders trying to breach the perimeter of the facility and alert the security guards 100 percent of the time.

- The PPS will identify vehicles and persons at the garage entrances and exits 100 percent of the time.

- The PPS will retain a video record of activity from all recorded cameras for at least 30 days and provide the ability to positively identify personnel captured in the images 100 percent of the time.

- The PPS will enable the protective force to remotely monitor and observe activity at all remote areas 100 percent of the time.

These objectives will help form the basis of the system design. They also provide the criteria for measuring the success of the project when the installation is complete.

System Design

The next step in the planning process is to decide who will design the system. There are two possibilities:

- **Customer or end user.** If the customer or end user knows exactly what functions the system is to perform, where all the cameras are to be installed, and where to place and how to operate the security control center, the system can be designed in-house. This is common where security managers and support vendors have determined the requirements for upgrading and operating the system. In these cases, the next step is to prepare the detailed specification for an invitation for bids (IFB).

- **Contractor or integrator.** Sometimes the customer or end user knows the problems to be solved but does not know how best to solve them. In this case, the normal approach is to describe the requirements and obtain proposals from several contractors or integrators. A problem is that the customer may not have sufficient knowledge to assess the solutions and costs submitted by vendors. This problem can be overcome with the help of a competent consultant.

Cost Estimates

The next step of the planning phase is to analyze the collected data and decide which physical security measures should be implemented. It may be necessary, for each alternative, to state its costs, benefits, and impact on business operations. The alternatives will then most likely be scrutinized by various operational units within the organization. Once a measure passes such scrutiny, the project can move to the design phase. In preparing cost and schedule estimates for physical security measures, the project team may want to use the services of a security consultant. It is also possible to obtain estimates from various security equipment vendors and use a composite. Chapter 6 offers guidelines on estimating security projects.

References

Walsh, Timothy J., and Richard J. Healy. (Eds.) *Protection of Assets Manual.* (Chapter 6.) Santa Monica, CA: Merritt Co.

CHAPTER 5:

DESIGN

Security Design Elements

A successful security design recognizes technological developments and integrates three primary elements: architectural aspects, security systems, and operational factors. Technology does not replace manpower but acts as a force multiplier to augment personnel capabilities and provide checks and balances to offset individual wrongdoing.

The human element of security, which relates to decision-making, common sense, and awareness, must be integrated into the system during the design phase, not after the system has been designed and implemented. Successful organizations realize that one of their greatest security resources is their employees. Only through a complete understanding of how employees will interface with the technology and respond to security incidents can companies prevent most security incidents and respond appropriately to those that occur. Operating procedures tailored to various threat levels must be developed before any systems are designed or purchased. An organization, whether governmental or private, that buys equipment and software without developing the operating philosophies presented in this book may end up with systems that actually increase the risk of harm.

The architectural aspect is one of the most significant factors in security design. Limiting the number of access points to a building and designing passageways through control points reduce the problems associated with controlling access. Good operating procedures must also be considered during the design phase. If not, the interface between man and machine will not be effective. Electronic security measures should be used to complement procedures. If everyone adheres to established, easy-to-follow procedures, abnormal activity becomes apparent.

Design Specification

One of the products of the design phase is the specification. If the procurement method is to use a request for proposals (RFP), the specification will be a functional specification—that is, one that spells out the system's desired functions and performance parameters. Vendors are requested to submit proposals showing how they will meet the requirements with hardware and software. If the procurement method is an invitation for bids (IFB), the specification will list specific equipment and software and request prices from the vendors to supply the specific items and install them. The customer might specify specific hardware and add the words "or equivalent" to allow some latitude in submitting the bids. Other IFBs may request a proposal from vendors if, for example, the customer has specified that a particular CCTV system be installed in a building and integrated with an existing access control system. The IFB would then request that bidders supply a written proposal on how they would achieve the integration.

Next it is time to prepare the statement of work, which details what work should be included in the contract. This document defines the removal of any old equipment and the supply, installation, and connection of new equipment and software. The document also contains instructions on coordination with other contractors; attendance at project meetings; testing of systems; commissioning activities; training of operators, systems managers, and maintenance personnel; warranty requirements; and follow-on maintenance requirements.

The design specification process is iterative. The team may repeat the steps and develop several drafts before proceeding to the next phase.

The three design elements—architecture, operations, and systems—all play a role. The interface between the users and security operations, as well as locations within the facility itself, must be evaluated before committing to particular security systems and device locations. With a new facility, much of this interaction has to be projected based on the drawings and is therefore more difficult.

An effective security design takes into account the following systems and functions:

Perimeter fencing and barriers
Intrusion detection at the perimeter
    and interior spaces
Site surveillance
Doors and locks
Key controls
Loading dock operations
Employee and visitor ID systems
Access control for employees
Visitor controls
Critical and restricted areas
Delivery controls and processing
Mail processing

Parking controls for employees and
    visitors
Stairwell ingress and egress
Elevator control
Interface with other systems, such as
    fire and building management
Critical site utilities
Guard posts and post orders
Security policies and procedures
Alarm assessment and dispatch procedures
Incident response procedures

Design Criteria

The design criteria help identify the characteristics that the security system will have in terms of performance and operational factors. The criteria also define the constraints, such as codes, standards, and costs.

System Performance

System performance refers to such items as response time to a card reader request or throughput of a revolving door or turnstile. Reliability and availability requirements may also be stated. The peak system load should incorporate at least a 50 percent expansion factor over the current situation. The peak loading profile for an access control system is the maximum number of access attempts per second to be handled systemwide. This rate depends on the type of access control device used (such as a card reader or biometric device) and the maximum number of devices used during peak periods. Based on simulation or actual tests, one can compute the time the system will take to process the peak number of access requests. For an efficient access control system, response time at any card reader should be no longer than two seconds at peak load. Response time is also important for database maintenance. Users should not have to wait longer than two seconds after a request to receive a response to a database update.

Capacity

This criterion should spell out minimum values for the size and space of various components of the system. It is also helpful to address expansion and spares requirements. For example, one should specify 25-50 percent expansion for disk files and memory above the maximum expanded configuration to allow for unforeseen applications and requirements that occur during the process. A rule of thumb for CPU computational power is that the computer should be able to execute intrusion detection programs, run access control programs, monitor the status of all equipment, accomplish database transactions, and produce at least four simultaneous reports within 5 to 10 minutes.

Special Features

The specification should avoid special features that will exclude some vendors. It is better to specify items that are commonly available, such as commercial, off-the-shelf (COTS) software. It may even be worthwhile to reengineer the organization's procedures to be able to incorporate an off-the-shelf system. If the system does not meet all management information or operational needs, it will need to be customized to the organization's environment. If the security department has a software staffer who can develop and maintain the system, that may be the best solution. Otherwise, the contractor must develop the system.

Any intrusion detection, access control, reporting, database maintenance, or monitoring functions that require development for the site should be developed in a high-level language, using the database for storage and retrieval of data. Machine-level language applications should be avoided. All documentation, source code, programming aids, and reference information should be delivered with the system. Otherwise, the company may be forced to use the integration contractor to update or change the application programs. The system should be delivered with the latest version of the industry standard operating system, and no modifications of the operating system should be required to run the software. Also, it may be wise to subscribe to an application program and operating system update service so that system will always have the latest software updates and documentation.

Codes and Standards

This criterion identifies any national or local building, fire/life safety, or other codes and standards that will be applicable to the project. It is also necessary to consider any company standards regarding security devices or operations.

Quality and Reliability

The design should specify reasonable levels of quality and avoid unnecessary items. The design should also pay attention to system reliability. Some deployments may obtain their maximum use during crises or disasters. However, such events may also impede the normal communications or power infrastructure used by the devices, hindering their reliability. The least expensive communications method may not be the most appropriate, depending on the needs for reliability.

Budget

The budget is developed as part of the planning process and constrains the system design to meet budgetary goals.

Critical Operations

Any constraints placed on the design by company operations must be defined and addressed.

Company Culture

Company culture always dictates certain requirements, especially aesthetic concerns. It is important that the security measures implemented complement and blend with the corporate image.

Monitoring and Response

Procedures for monitoring and responding to system alarms must be addressed in the design phase, not during the installation phase.

Maintenance

Maintenance of the hardware and software should be addressed at the design level to ensure that the system is easily maintained. With technology advancing quickly, it is important that the system have the most recent releases of hardware and software products from the manufacturer. This may require that systems applications development be done on an earlier release of hardware and software, with a migration to the new equipment and a new release of the operating system just prior to installation. Using this approach, the customer avoids taking delivery of an obsolete system.

The design should keep future maintenance in mind. One technique is to use life-cycle cost analysis in the procurement process. Further, technologies should be deployed to reduce the frequency and inconvenience of maintenance activities. It is wise to use well-known products that have proven reliability, modular and standardized components, self-diagnostic and self-healing technologies, and technologies that improve repair access (for example, CCTV cameras that may be lowered mechanically on their tower, obviating the need for a bucket truck). These technologies may require spending more money at the inception of the project but should result in lower maintenance costs during the operations phase of the project.

Products of the Design

The output of the design phase is termed the "bid package" or "construction documents." It consists of bidders' instructions, specifications, drawings and schedules, and hardware schedules.

Bidders' Instructions

This document describes the company's requirements for qualification of the bidders, such as licenses, labor affiliations, experience, and bonds. Formats for those various documents are included, as are the evaluation criteria and weights. The time frames for each required activity are also delineated.

Specifications

The design specifications document the security systems requirements in sufficient detail that potential bidders get a common understanding of the functions required of the system and what has to be
done to install it. The Construction Specifications Institute (CSI) has
established a standard format for construction projects. The CSI Mas-
terFormat™ currently has 16 divisions with security requirements being defined under at least five different sections. The CSI MasterFor-
mat™ is currently being modified with a planned release in 2004 that
has a separate section for electronic security systems. Because most
security systems are integrated systems and are procured through a
single integration contractor, many system designers are departing
from the standard format and including a single section for all security systems requirements. Within each division, the subsections are
presented in a standard format, consisting of three parts: general,
products, and execution. These parts may be further subdivided as
needed to provide the necessary clarity for the specification.

Drawings and Schedules

A very important section of the bid package is the drawings section.
Security systems require many types of drawings and schedules. Special security symbols developed by the Security Industry Association
have been adopted by American Society for Testing and Materials
International.

The most commonly supplied drawings are the following:

- **Plan drawings.** A plan drawing shows an area in map-like
  form to specify where at a particular site the security devices
  are located.

- **Elevation drawings.** Elevations are drawings of vertical sur-
  faces to show how security devices are mounted on a wall or in
  racks.

- **Details drawings.** Details show cable terminations or special
  mounting requirements.

- **Riser diagrams.** These diagrams show complete security subsystems, including all the devices and how they are connected
  in a building or campus.

- **Conduit and cable lists.** These lists show the various types and quantities of conduit and cables.

Hardware Schedules

Several schedules are included to aid in understanding the specific components to be provided:

- **Data panel schedule.** This schedule lists the types, locations, and communications aspects of the various panels used throughout the facility.

- **Door hardware schedule.** This schedule lists all doors in the facility and notes their type, location, device used for access (such as card reader), type of lock, method of egress, and other information that describes how the doors function.

- **CCTV camera schedule.** This schedule lists all the security cameras in the system, showing camera number, location and view, type of camera, type of lens, type of enclosure, mounting method, and any alarm interface.

Operating Procedures

Before finalizing the design of the PPS and selecting hardware and software, it is prudent to develop the procedures that will be used to assess the alarms of the PPS subsystems and to respond to the different types of intrusions.

Alarm Assessment and Dispatch Procedures

Access control and intrusion alarm systems are designed to provide notice if an intruder tries to enter a facility. For example, an access control system should alarm if an unauthorized person tailgates behind an authorized person into a facility; an intrusion detection system should alarm if a person opens a door or window at the wrong time or without presenting a valid ID card or code. For each alarm point, a facility should have detailed operating procedures for the person assigned to monitor the alarms. The following table shows the type of information that should be recorded for each alarm point. The shunt column indicates the action required to prevent an alarm at each control point.

Table 5-1
Alarm Point Definition

Location	Time Activated	Sensor Type	Alarm Type	Shunt	CCTV Interface	Assessment Procedure
Front Door	1800-0600	Door contact	Forced open	Valid card read	Pre/post alarm recording of camera #1 scene	Front door #1
Front Door	1800-0600	Door contact	Held open		Pre/post alarm recording of camera #1 scene	Front door #2
Computer Room	24/7	Motion detector	Unauthorized entry	Valid card read	Pre/post alarm recording of camera #2 scene	Computer room #1
Loading Dock	24/7	Duress button	Duress		Pre/post alarm recording of camera # 3 scene	

In addition, it is important to document the location and purpose of each CCTV camera and any interfaces it may have with the intrusion detection or access control subsystems. Table 5-2 shows the type of information to document.

Table 5-2
CCTV Camera Definition

Camera	Location	Purpose	Special Features	Time Activated	Triggers
#1	Inside front door looking out	To identify personnel entering front door	Backlighting adjustment; person's face must take up 20% of camera view	1800-0600	Door forced open alarm
#2	Inside computer room looking toward door	To identify personnel entering computer room	Backlighting adjustment; person's face must take up 20% of camera view	1800-0600	Motion detector alarm in computer room

The systems integrator can use the information in tables 5-1 and 5-2 to program the PPS to provide the alarms specified and to trigger the CCTV cameras and recorders to display information for the person assigned to monitor the systems.

The next step is to write detailed procedures that will enable the operator in the security control center to correctly assess the alarms. The procedures should be displayed by the system so the operator can follow them immediately. An indexed book of procedures for each alarm should also be available. Table 5-3 shows an example of a procedure for a "door forced open" alarm at the front door of the facility.

Table 5-3
Alarm Assessment Procedures

Assessment Procedure ID #	Time	Alarm Type	Assessment Actions
Front Door AP #1	1800–0600, 5 days per week; 24 hrs Saturday and Sunday	Door forced open	View current image of camera #1 on monitor #1. View pre-alarm image from camera #1 on monitor #2. If suspicious activity, dispatch patrol officer to location. Acknowledge alarm on access control system. Enter info in "action taken" window.

Response Procedures

The next step is to prepare procedures for the security force that will respond to alarms. A procedure must be developed for each type of alarm and each type of scenario that might be encountered by the responding officer. It takes a great deal of thought and imagination to list all the different types of scenarios an officer might encounter. Nevertheless, it is important to list as many scenarios as possible. Table 5-4 shows an example of an incident response procedure.

Table 5-4
Incident Response Procedures

Incident Response Procedure ID #	Time	Alarm Type	Scenario	Response Actions
Front door IRP #1	1900–0700, 5 days per week, 24 hrs Saturday and Sunday	Door forced open	Door opened; invalid ID card read	Approach intruder and challenge. Request identification. Radio security control center with information. Follow direction from security control center.

Once all the necessary information has been compiled, the security manager will have a good understanding of the security system and how it will help the security force protect the facility. With the information, a security systems integrator will be in a much better position to install a system that will be usable by the personnel operating it.

Form of Specification

The form of the specification depends on the type of procurement and the technical knowledge of the group producing it. This may be a case where a little knowledge can be very dangerous and costly; one should not underestimate the technical knowledge required. The specification should make it easy for bidders to understand exactly what they are expected to supply and install without wading through reams of documents. The more difficult it is to understand a specification, the more variation there will be in bidder proposals and prices.

A typical specification may be broken down into the following headings. Suggestions are provided for some specific items that may be overlooked. Most companies and local authorities produce impressive and voluminous contract conditions to protect themselves. The specification should adhere to the standard three sections suggested by the CSI format: general, products, and execution. The example that follows includes sample, additional subsections.

Model Specification

Part 1: General

Authority and Responsibility

In this section, state who is issuing the specification, who has responsibility for making any changes, and who should be contacted and in what manner for any questions and comments.

Summary

This section contains the following:

- Overall project description: high-level description of the overall project if PPS is part of a larger construction project

- List of all documents included in the bid package

- PPS description: a high-level, general description of the system

- System operation: a brief description of how the system will be operated

- Description of all products and services to be included in the contract, including the supply, installation, and connection of PPS components and cable

Objectives

This section lists the system objectives so all bidders can understand what the system is intended to achieve. The objectives should be SMART.

Submittal Format

Here the customer describes the outline and format for the proposals and specifies all the items to be included in the submittal. This section should also specify all system options that should be priced separately. The evaluation process, evaluation criteria, and criteria weighting are also explained in this section. Clear instructions in this section will greatly simplify the proposal evaluation process.

Performance Specifications

When using the RFP method of procurement, the specification is given as functional performance requirements for the system and equipment. It is the responsibility of the bidders to select the most appropriate equipment to fulfill the objectives and requirements of the system. Certain items may be specified by manufacturer and model number when necessary to ensure compatibility or perform-ance. Part of the proposal evaluation process is to assess the quality, reliability, and suitability of the equipment proposed.

Future Expansion

This section describes any capacity, capability, or performance ex-pansion requirements.

System Interfaces

Here the customer indicates whether other systems will or might be connected or interfaced to this system.

Codes and Regulations

The installation should comply with all relevant regulations, such as the following:

ADA: Americans with Disabilities Act
ASCII: American Standard Code for Information Interchange
ASTM: American Society for Testing and Materials
EIA: Electronic Industries Alliance
FCC: Federal Communications Com-mission

NEC: National Electrical Code
NEMA: National Electrical Manufac-turers' Association
NFPA: National Fire Protection Asso-ciation
UL: Underwriters Laboratories, Inc.

Customer Supplied Materials and Services

This section lists any items provided by the customer, such as storage facilities, power supplies, or tools.

Scheduling

Here the customer states the likely time frame for contract placement and job completion.

Statement of Compliance

All bidders must include a statement that the system proposed and priced complies with the specification. Variations and suggestions for changing or improving the system should be listed and priced separately.

Indemnity and Insurance

The contractor should indemnify and keep indemnified the customer against injury to, or death of, any person and loss of, or damage to, any property arising out of or in consequence of the contractor's obligations under the contract and against all actions, claims, demands, proceedings, damages, costs, charges, and expenses in respect thereof. For all claims against which the contractor is required to insure, the insurance coverage should be a minimum of $1 million or such greater sum as the contractor may choose in respect of any one incident. The contractor should be expected to produce evidence of sufficient insurance coverage to meet these requirements before any work is carried out on-site.

Bonds

A surety bond is a three-party instrument between a surety (or insurance company), the contractor, and the project owner or customer. The agreement binds the contractor to comply with the terms and conditions of a contract. If the contractor is unable to successfully perform the contract, the surety assumes the contractor's responsibilities and ensures that the project is completed. Below are the four types of contract bonds that may be required:

- **Bid.** This type of bond guarantees that the bidder on a contract will enter into the contract and furnish the required payment and performance bonds.

- **Payment.** This type of bond guarantees payment from the contractor to persons who furnish labor, materials, equipment, or supplies for use in the performance of the contract.

- **Performance.** This type of bond guarantees that the contractor will perform the contract in accordance with its terms.

- **Ancillary.** These are bonds that are incidental and essential to the performance of the contract.

Modifications and Variations

No modifications or variations to the contract should be permitted without the written consent of the customer.

Notification

The contractor should notify the customer immediately if any unforeseen circumstances are encountered during the course of the contract that may require modifications or variation. The contractor should then await instructions before proceeding with any part of the contract that may be affected.

Warranty

The contractor should be required to repair, correct, or replace any defect of any nature that may occur for a period of 12-24 months from the date of issue of the certificate of practical completion. The common time for the contractor to report to the job site to address a warranty issue is within four hours of the problem report. The problem should be corrected without undue delay. Therefore, the contractor should hold sufficient stock of spares to allow speedy repair or replacement of equipment. Waiting for manufacturers to replace or repair equipment is not acceptable. The contractor should provide the employer with details of telephone and fax facilities for reporting all problems and defects. The warranty should include full maintenance of equipment in accordance with the manufacturer's recommendations. The contractor should have in operation a system whereby all service visits are recorded in a database, and a report form should be provided to the customer. The report form should record the date and time that the fault was reported, the nature of the reported fault, the date and time of the visit, the actual fault identified, and the remedial work carried out.

Maintenance

The contractor should submit a full schedule of maintenance to be carried out on the system during the warranty period and under subsequent maintenance contracts. This section should also state that the contractor must install all hardware and software updates and upgrades that become available during the time the system is being installed and is under warranty. This provision is included to protect the customer from having to accept a system that is obsolete upon installation.

Part 2: Products

This part of the specification lists equipment. One approach is to specify every item by manufacturer and model number. The advantage of that approach is that a totally objective comparison of all bids can be made. The disadvantage is that such specificity precludes the use of other, possibly less expensive or perhaps better makes of security devices that have performance characteristics as good as or better than those specified. By specifying one model, the customer provides an advantage to the company that has the best terms with that particular manufacturer. An alternative approach is to produce a performance-related specification with generic device descriptions. However, that approach requires especially careful bid assessment. Generally, a performance specification leads to the most competitive prices. Another commonly used technique is to specify a manufacturer and model number but follow it with the words "or equal."

The following are some of the product categories to list in this section:

Card readers	Matrix switchers
Access control panels	Telemetry receivers
ID cards	Quad units
Workstations	Video printers
Transmission of video and telemetry	Consoles
Cameras	Monitors
Lenses	Cabling
Distribution amplifiers	Power supplies
Monitors	Enclosures
Camera housings	Intercom equipment
Pan, tilt, and zoom units	Duress buttons
Video recorders	Motion detectors
Multiplexers	Door contacts

Non-Proprietary Equipment

This section should state that all equipment, consoles, telemetry, switching and multiplexing devices, and other hardware must be commercially available, off-the-shelf products. This requirement ensures that future extensions to the system may be carried out by any installing company. The use of specialized, in-house manufactured components should not be allowed unless specifically requested as part of the requirements.

Part 3: Execution

Preparation of Site

Here the customer describes the condition of the site where the system will be installed and the work to be done by the contractor to prepare the site for the new system.

Installation and Quality Control Standards

This section states how inspections and quality control procedures will be conducted and records will be kept.

Trade Coordination

This section states whether coordination is required with other contractors regarding, for instance, fiber-optic cable installation or local-area network (LAN) or wide-area network (WAN) connectivity.

Subcontracting

No part of the contract should be subcontracted to any other company or individual without the express written permission of the customer. Unless specified to the contrary, it is assumed that all work will be carried out by the contractor. If the contractor intends to subcontract any part of the design or installation, that intention must be made clear in the bid submission and the name of the subcontractor should be provided. The customer should reserve the right to accept or reject nominated subcontractors.

Special Equipment

The contractor should normally be responsible for providing all special equipment necessary for safe installation of all high-level equipment. It should be the contractor's responsibility to provide all access equipment required to complete the installation in accordance with good safety practices.

Health and Safety

The contractor should be expected to comply with all health and safety requirements of the customer and of the authority having jurisdiction (AHJ).

Preassembly and Testing

All equipment should be prebuilt and tested at the contractor's premises before being delivered to the facility. The telemetry controls, multiplexer controls, and central VCR time/date generation should be assembled and proved to the satisfaction of the client's representative before being delivered to the site.

Testing and Commissioning

When the contract is considered to be complete, a certificate of completion should be issued after successful completion of reliability testing.

Operating Instructions

The contractor should provide a minimum of four full sets of operation manuals, operating instructions, descriptive brochures, and technical manuals for all subsystems included in the contract.

As-Built Drawings

The contract should require the contractor to provide as-built wiring and schematic diagrams.

Training

This section specifies what training will be required, over what period, and where. It also indicates what training manuals should be supplied and in what media. This section also states what qualifications (such as certifications from manufacturers) are required of the trainers.

Programming

Here the customer should ask the contractor to submit a list of all proposed programming activities, including device names, descriptions, timing, and sequence of operations. This section should also specify all programming to be done by the contractor for all subsystems.

Upgrades

The contractor should provide and install all hardware and software upgrades that become available for the PPS during the warranty period at no additional cost.

References

Walsh, Timothy J., and Richard J. Healy. (Eds.) *Protection of Assets Manual.* (Chapter 6.) Santa Monica, CA: Merritt Co.

CHAPTER 6:

ESTIMATION

Preparing cost estimates of the various physical protection systems under consideration greatly helps in understanding the components and costs of each system configuration. It also establishes a range of costs and sets limits for procurement. Often, upper management will require estimates and a comparison of alternatives before the project can proceed.

Types of Cost Estimates

Several types of cost estimates are used in the implementation of physical protection systems: budgetary estimates, preliminary design estimates, and final design estimates.

Budgetary Estimates

Budgetary estimates are prepared during the initial planning phase for a new PPS. The goal is to arrive at a cost figure that can be used for getting the new PPS into the budget cycle. Depending on the company's procurement policies, the budget cycle may require that systems be identified and submitted for consideration several (as many as five) years before the planned implementation. Since these estimates are used for budgetary purposes, they have a large contingency, such as plus or minus 10 to 20 percent. These estimates are difficult to prepare without actually performing a good portion of the system design.

To prepare a budgetary estimate, the project manager can discuss costs with other companies that have recently installed systems and ask potential vendors to develop budgetary estimates. Another resource is the data developed by RSMeans, a company that provides construction cost information.

Preliminary Design Estimates

If the PPS project is part of a larger construction project, the process may require a preliminary design estimate. This estimate should be developed at the 50 percent design review stage and normally has a contingency of plus or minus 10 percent. Since the design of the system is well under way, draft specifications, drawings, and equipment schedules can be used to develop the costs. Potential vendors, too, can provide estimates.

Final Design Estimates

The estimate is refined as the project advances to 100 percent completion. At this point, the final design estimate is developed using the completed documents, drawings, and schedules. This estimate should have minimal contingency, on the order of plus or minus 5 percent.

Life-Cycle Cost

The actual cost of a PPS is its life-cycle cost. Life-cycle cost estimates include the following components:

- **Engineering and design costs.** These are the costs associated with the design of the PPS, such as determining the appropriate products to accomplish the functions specified and producing drawings showing equipment locations, subsystem connections, and details of wiring various devices.

    — **Hardware.** The hardware costs include the original equipment plus startup spare parts.

    — **Software.** The software costs are for the operating system and application system software necessary to operate the PPS.

- **Installation costs.** Installation costs include labor expended in installing equipment and software, labor to perform inspection, testing and commissioning, equipment rental, permits, bonding, supervision, and overhead.

- **Operating costs.** Operating costs include expenses for personnel, power consumption, and consumables (such as paper and ink cartridges).

- **Maintenance costs.** Maintenance costs include labor and spare parts for preventive and remedial maintenance.

- **Other costs.** Other costs include state and local taxes, profit (10 percent), performance bonding (3-5 percent) and contingency (5-10 percent).

- **Adjustments.** RSMeans data are based on national averages. For specific locations, the cost data may need to be adjusted.

Detailed Estimating Procedures

The following is a step-by-step process for preparing an estimate for a physical protection system:

- **Identify PPS subsystems.** Typical subsystems for a PPS include the following:

— **Fences and barriers:** perimeter fences; enclosures or fences around critical utilities; and portable, fixed, and automatic barriers

— **Security control center or monitoring subsystem:** consoles, workstations, computers, printers, recorders, and displays

— **Access control subsystem:** card readers, badges, badge preparation equipment, door locking devices, door position sensing devices, turnstiles, drop gates, and mantraps

— **CCTV subsystem:** cameras, switchers, recorders, mounts, enclosures, and monitors

— **Intrusion detection interior and exterior subsystem:** intrusion detection sensors, alarm sounding devices, and display devices

— **Lighting:** lighting fixtures, mounts, enclosures, poles, and ballast

— **Power, control, and data distribution:** backup power, surge protection, raceways, grounding, conduit, wire, and cable

— **Communications subsystem:** communications network used to connect all subsystems, intercom, radio, network, and telephone equipment

— **Search equipment:** metal detectors, explosives detectors, and X-ray machines

- **Identify other installation activities.** Before PPS components can be installed, the site must be prepared as follows:

— **Site civil or structural modifications:** grading, drainage, towers, foundations, fencing gates, and barriers

— **Specialty construction:** guard houses, monitoring stations, and ballistic- and blast-resistant structures

- **Develop list of components for each subsystem.** This information can be obtained from equipment vendors' brochures and guidelines or from RSMeans publications.

- **Establish component prices.** Cost data can be obtained from vendors and from RSMeans.

- **Estimate installation labor.** System integrators and equipment manufacturers can provide information regarding how many personnel and how much time will be required to install each component. They can also report the normal hourly rates.

- **Identify required special equipment and rates.** For the specialty construction activities required to install a PPS, one must identify, for each activity, the number of personnel and hours required, any special equipment needed, and the rental cost of that equipment.

- **Use spreadsheet program.** Once all the information has been gathered, the project manager should construct a spreadsheet to compute the estimate for the project.

In summary, it is important to use actual cost data or recent quotes from vendors whenever possible. The RSMeans industry averages are useful, but they must be adjusted to specific locations. Also, because estimates contain some contingency, the project manager should expect variation when bids are received.

In addition, a quality review is essential. After gathering the data and preparing the estimate, the project manager should subject them to a comprehensive review process to ensure that all the components are listed and in the correct quantities. The review should also double-check the cost of labor and the number of personnel required for installing each component, make sure there have not been any recent price increases, and determine whether any ongoing or near-term changes at the site may affect the project.

Sample Estimate

Table 6-1 shows a sample spreadsheet for an integrated PPS. The data for this spreadsheet were received from security equipment vendors in the San Francisco area. The sample security system consists of the following components:

- 2 perimeter revolving doors

- 10 interior single-leaf doors

- 12 fixed CCTV cameras

- 1 pan, tilt, and zoom camera on the roof of the building

- 2 CCTV monitors

- 1 digital video recorder

- 1 computer monitor for access control

Table 6-1
Sample Estimate

QTY	Description	Unit Cost	Extension	Labor
1	Revolving door	$30,000.00	$30,000.00	$15,000.00
2	8 card reader control panel	$1,000.00	$2,000.00	$1,000.00
1	Access control software	$1,000.00	$1,000.00	$500.00
11	Proximity card reader	$100.00	$1,100.00	$1,100.00
3	12 VDC batteries	$20.00	$60.00	
10	Request to Exit Motion Detectors	$75.00	$750.00	$1,000.00
1	24 VDC power supply	$87.00	$87.00	$100.00
10	24 VDC electric strikes	$150.00	$1,500.00	$2,000.00
1	Operator training			$500.00
1	PTZ CCTV camera, lens, and enclosure	$2,500.00	$2,500.00	$2,500.00
12	Fixed CCTV camera, varifocal lens, and enclosure	$700.00	$8,400.00	$8,400.00
Subtotals			$47,397.00	$32,100.00

Summary of Costs				
Components		$47,397.00		
Tax & Shipping (12%)		$5,687.64		
Installation Labor ($100 per hour)			$32,100.00	
Subtotal				$85,184.64
Profit (10%)				$8,518.46
Subtotal				$93,703.10
Performance Bond (4%)				$3,748.12
Total Cost Estimate				$97,451.23

References

Owen, David D. (2003). *Building security: Strategies & costs.* Kingston, MA: Construction Publishers and Consultants.

CHAPTER 7:

PROCUREMENT

Procurement Approaches

Clearly, before an organization can use a new physical protection system, it must purchase—or procure—the system. The three common methods of procurement are as follows:

- **Sole source.** This type of procurement involves only one vendor and is used when the customer has intimate knowledge of the requirements defined and the systems available on the market. In some cases, the vendor has the only product on the market that can fulfill the organization's requirements. The customer and the vendor then enter into negotiations on the costs for time and materials.

- **Request for proposals.** An RFP is the most common method of procurement in the security field. The customer issues an RFP containing functional requirements of the security system. Prospective vendors prepare proposals regarding how they will meet the requirements, spelling out the equipment and software they will supply, installation methods, and a cost breakdown. The customer then evaluates each proposal against pre-established criteria and awards the contract to the winning vendor.

- **Invitation for bids.** An IFB procurement is used when the customer has completed its own security system design, including the selection of hardware and software. The invitation for bids asks vendors to state the cost of materials and installation. Usually, the qualified bidder with the lowest price is awarded the contract.

Sole Source

In a sole source procurement, the customer selects a single vendor or contractor, negotiates the equipment and installation costs, and works with the contractor to design and install the system. This method is recommended for small projects and for upgrading existing systems. The advantage of this method is that the planning and design phases are simplified and shortened, thus saving design costs and reducing the time to complete the project. To ensure fairness, regulations may prohibit the use of sole source procurement.

Request for Proposals

Government agencies were among the first to use the RFP as a way to determine the lowest bid for products. Companies, particularly those required to accept the lowest bid, quickly adopted the RFP process to itemize the specifications desired in a system. Companies continue to justify the RFP process as a way to assess requirements, involve staff, negotiate a contract with a vendor, and appear fair when seeking products and services from competing vendors. RFPs are costly in terms of time and money for all parties involved. Using an RFP, by the time a company selects a vendor for its new security system, 18 to 24 months may have passed. By then, the technology specified in the RFP may already be outdated.

Despite the serious problems inherent in the security system RFP process, it is very unlikely to be abandoned soon. Three ways to streamline the RFP document, and by extension the responses, are as follows:

- Provide an outline for the bidders to follow in their responses, giving them an explanation of what the purchaser wants in each section.

- Provide clear instructions for the proposal content. To test whether the instructions are clear, the project manager can show them to a colleague who knows nothing about the project. If he or she understands them, the instructions are probably clear.

- Limit the length of each section of the RFP, including vendor responses. For example, it might make sense to allow two pages for stating the requirements for each function. Generally, a 20- to 30-page RFP generates about twice as many pages in the response, plus diagrams, supporting documentation, and sample reports. To minimize the number of pages that must be reviewed, the project manager should consider why the RFP asks vendors for explanations and what is expected in response.

Typically, an RFP contains at least the following elements:

- Contact names and information sources for the RFP and for the contract

- Background on the company and the security environment

- Results of the requirements assessment

- Schedule of key dates, including when the response is due, when the decision will be made, when funds will be available, and when installation is expected to begin

- Instructions for formatting the response, stating what the vendor may or may not include in the response

- Specific requirements, grouped by function or other commonality

- Operating environment requirements, such as the operating system or network environment

- List of documents required as attachments, such as sample reports, drawings, and schedules

Providing the RFP document in electronic format will help vendors expedite their response, as they will not have to scan or retype the document. It is helpful to use a common word-processing program, such as Microsoft Word, and to make the document available on diskette or on-line for downloading.

Vendors should be allowed to submit questions for clarification. They want to ensure that they understand the purchaser's requirements, culture, and environment. Some RFPs, for example, contain terminology that reflects an old system, rather than current technology. Allowing questions up front reduces the company's risk of being misunderstood and the vendors' risk providing misinformation. All vendors should receive all questions and answers.

Bidders' Conference

The project manager should schedule a bidders' conference about a week after the RFP is issued. The conference enables vendors to see the facility, view any possible installation problems, and ask questions about the design requirements. A bidders' conference brings all potential bidders together in a room so the project manager can explain the procurement requirements to all of them at the same time. The project manager can also use the occasion to provide further insight into what the organization is trying to accomplish with the project and to enable the potential contractors to see each other. The conference thereby establishes that the project manager is in control of the procurement. It also strengthens the purchaser's negotiating power by showing potential bidders that they have competition.

To be successful, a bidders' conference must be organized and executed with precision. It should begin with opening remarks from the project team leader, who should introduce the evaluation team. Each bidder should introduce himself or herself to the group. The project team leader should present the agenda, identify the conference guidelines, and present an overview that demonstrates how the project will match corporate strategy and states the key objectives to be met.

Next a member of the evaluation team should discuss the procurement process, identifying all steps in the procurement, including key dates. This is the time to stress key points—for example, that the successful bidder must commit contractually to deliver a complete solution, assume responsibility, share risks, and deliver on time and within budget. The project manager and technical team members should discuss the project's purpose, the results to be produced, the general time line, the evaluation criteria, and any relevant technological or architectural issues. Suppliers should then be given an opportunity to tour the site and ask questions. It is wise to record all the questions and send them and the answers out to bidders. An alternative is to require that all vendors' questions be submitted in writing after the conference and then send answers to all vendors represented. Such an approach gives an appearance of formality, fairness, and equality of information, which stimulates competition.

Dealing with all potential bidders at the same time saves much time and effort. A properly run bidders' conference may take two to four hours and add value and power to the procurement process, resulting in better proposals. However, bidders' conferences may have no value for either side if the project is simple and there is not much work to prepare the site. Some vendors may have to send representatives from out-of-town, and the attendant expenses may needlessly raise the cost of doing business.

Invitation for Bids

Sometimes companies have the internal expertise to design their own systems; other times they hire a security consultant to design their systems. When the system has been completely designed, the design document can be sent to several qualified vendors, and they can be asked to bid on supplying and installing the equipment.

The purpose of the IFB is to obtain the lowest cost for the equipment, software, and installation. No technical proposal is requested, and pricing is provided in the format requested. This method requires an up-front effort to develop the detailed design but shortens the bid preparation time.

There are many variations on this procurement technique. For instance, the IFB may ask for specific quotes on supplying and installing security devices such as CCTV cameras, but it may also ask for a proposal on how to integrate the devices with an existing access control system. In such a case, the procurement evaluation process is much more difficult, as the successful bidder must not only have the lowest price for the components to be installed but also present the best technical proposal for the systems integration and the lowest price for all activities.

Evaluation Criteria

The purchasing organization should be clear about its evaluation criteria, determining the criteria, how much weight each criterion has, and the method for evaluating proposals. Settling this early provides some framework for the format of the RFP, if that procurement approach is used. After the document is issued, the organization should not change the criteria unless it is prepared to issue a complete revision. Piecemeal amendments can compromise the whole project.

The following are some questions to consider in evaluating the proposals:

- Does the proposal carefully describe the project's objectives and importance, as well as the viability and appropriateness of the chosen solution?

- Is the approach scientifically and technically feasible?

- Is there a justification of the proposal's underlying theories, methods, and techniques?

- What are the qualifications of the engineers and technicians participating in the project?

- Does the contractor provide proven capabilities through other projects?

- Does the proposal clearly specify intermediate and final project results, deliverables, and products?

- Are there established milestones and reporting schedules for the project's progress and success?

- Does the proposal allocate manpower realistically throughout the project?

- Does the proposal justify the use of major existing and new equipment, supplies, and other items?

- Will the solution reduce long-term cost or duplication?

- Can the functional security requirements be met, given the proposed architecture?

- How does the proposal address assurance in integration of all subsystems?

- Does the solution properly include security operating procedures and practices?

- Does the proposal address physical protection system life cycle support?

Table 7-1 shows sample criteria that could be used for evaluating vendor proposals.

Table 7-1
Sample Evaluation Criteria

Criteria	Available Points	Actual Points
A. Proposed Technical Solution • Components and peripherals • Display and assessment • Control • Communication network • Software development and support	600	
B. Implementation Plan • Installation methods and schedule • Testing • Training	300	
C. Maintenance and Support • Staff • On-call services • Equipment maintenance and supplies	100	
D. Experience • Project staff • Similar projects	100	

Criteria	Available Points	Actual Points
E. Background and Financial Viability • Three years' financial statements • Disclosure of litigation • Prevailing wage participation	100	
F. Cost • Hardware cost • Software cost • Maintenance cost (after warranty) • Operating cost (personnel, spare parts, etc.)	600	
G. Total	1,800	

Selecting Vendors

One of the most important steps in the procurement phase is to include a good selection of capable and competent suppliers and contractors. For this reason, many companies pre-qualify their vendors before sending them an RFP or IFB. This step requires the same amount of effort as hiring a new employee. Qualifications must be reviewed, references checked, and interviews held with the people who will perform the work. It is also important to ensure that the personnel identified in proposals are the ones who will actually perform the work. The purchaser should ask for written employee guarantees from the contractor. Once three or four qualified contractors have been identified, the proposal can be sent to them alone. Such an approach saves time compared to sending the RFP or IFB to all available vendors.

Bidder Qualifications

The following bidder qualifications were taken from a recent invitation for bids by a government agency. The installer/dealer had to provide proof of the following:

- State contractor license of C-7 or greater

- ACO (authorized central office) license

- Authorized dealer of the proposed manufacturer for a minimum of three years

- Minimum of five references (including name of client, address, system description, and contact person's name and telephone number) for prior security systems installations of the proposed manufacturer's equipment

- General liability insurance of at least $2 million

- Commercial vehicle insurance of at least $1 million

- Ability to provide a performance bond if requested

- Approval by Underwriters Laboratories, Factory Mutual, and state fire marshal (if a central station monitoring company)

Quality Assurance

The following requirements should be considered for quality assurance:

- All equipment supplied shall be listed by a nationally recognized test laboratory where applicable.

- All equipment and accessories shall be the product of a manufacturer regularly engaged in its manufacture.

- All items of a given type shall be the products of the same manufacturer.

- All items shall be of the latest technology; no discontinued models or products are acceptable.

- The manufacturer, or its authorized representative, shall confirm that within 20 miles of the project site there is an established agency that

  — stocks a full complement of parts,

  — offers service during normal working hours as well as emergency service on all equipment to be furnished, and

— will supply parts and service without delay and at reasonable costs.

- The contractor or authorized dealer shall be capable of performing service or maintenance work on these specified or accepted systems. The contractor shall be factory-certified where such certification is available.

Awarding the Contract

Regardless of whether an RFP or IFB is used, the project team evaluates the vendors based on criteria established during the design phase. When comparing costs, the team should take care to use the life cycle cost, which is the purchase cost plus the maintenance cost over the useful life of the system. It is also wise to check the financial stability of the vendor and specific customer references.

After evaluating the proposals, many organizations narrow the selection down to two or three vendors. They then interview vendor staff and ask about their experiences on other projects. These interviews provide valuable insight into the contractor's understanding of the project and ability to complete the project successfully.

Chapter 8:

Installation and Operation

At this stage, the project manager should instruct the contractor on installing all system components, including any customer-furnished equipment. The contractor must install all subsystems in accordance with the manufacturer's instructions and any pertinent installation standards. The contractor should furnish all necessary connectors, terminators, interconnections, services, and adjustments required for a complete and operable system. After installation, the project manager can tune the system to the specific operations of the facility.

Planning the Installation

The most important step in installing the PPS is to plan correctly. All the door hardware, card readers, sensors, panels, cameras, monitors, and console equipment should already be included in the design package and located on drawings. The installation contractor should verify the locations and note any changes needed. Together, the project manager and installation contractor should examine the installation requirements and make sure all issues and differences have been resolved before proceeding.

Next, the contractor should visit the site and verify that conditions agree with the design package. The contractor should be required to prepare a written report of all changes to the site or conditions that will affect performance of the system. Also, the contractor should be instructed not to take any corrective action without written permission from the customer.

It is also important that the contractor inspect, test, and document all existing physical protection equipment and signal lines that will be incorporated into the new system. For nonfunctioning items, the contractor should provide specification sheets or written functional requirements to support the findings and should note the estimated cost to correct any deficiencies. Also, in the report the contractor should note the scheduled date for connection to existing equipment. The contractor should not disconnect any signal lines or equipment or create any equipment downtime without prior written approval of the customer. If any device, signal, or control line fails after the contractor has commenced work on it, the contractor should diagnose the failure and correct it. The contractor should be held responsible for repair costs due to negligence or abuse of the customer's equipment.

Component Installation

Details on installing PPS components can be found in a new draft standard from the National Fire Protection Association (NFPA 731). General installation considerations are given in the following sections.

Card Readers

Card readers should be suitable for surface, semi-flush, pedestal, or weatherproof mounting as required. They should be installed in accordance with local codes, the requirements of the authority having jurisdiction (AHJ), and any other applicable local, state, or federal standards.

Electric Door Strikes or Bolts

Electric door strikes or bolts should be designed to release automatically (fail safe) or remain secure (fail secure), depending on the application, in case of power failure. They should use direct current (DC) to energize the solenoids. Electric strikes or bolts should incorporate end-of-line resistors to facilitate line supervision by the system. The following are some other installation considerations:

- **Solenoids.** The actuating solenoid for the strikes or bolts should not dissipate more than 12 watts and should operate on 12 or 24 volts DC. The inrush current should not exceed 1 ampere, and the holding current should not be greater than 500 milliamperes. The actuating solenoid should move from the fully secure to fully open positions in not more than 500 milliseconds.

- **Signal switches.** The strikes or bolts should include signal switches to indicate to the system when the bolt is not engaged or the strike mechanism is unlocked. The signal switches should report a forced entry to the system.

- **Tamper resistance.** Electric strike or bolt mechanisms should be encased in hardened guard barriers to deter forced entry.

- **Size and weight.** Electric strikes or bolts should be compatible with standard door frame preparations.

- **Mounting method.** The electric strikes or bolts should be suitable for use with single and double doors with mortise or rim hardware and should be compatible with right- or left-hand mounting.

Electromagnetic Locks

Electromagnetic locks should contain no moving parts and should depend solely on electromagnetism to secure a portal, generating at least 1,200 pounds of holding force. An electromagnetic lock should release automatically in case of power failure. It should interface with the local processors without external, internal, or functional alteration of the local processor. The electromagnetic lock should incorporate an end-of-line resistor to facilitate line supervision by the system. The following are some other considerations:

- **Armature.** The electromagnetic lock should contain internal circuitry to eliminate residual magnetism and inductive kickback. The actuating armature should operate on 12 or 24 volts DC and should not dissipate more than 12 watts. The holding current should be not greater than 500 milliamperes. The actuating armature should take not more than 300 milliseconds to change the status of the lock from fully secure to fully open or fully open to fully secure.

- **Tamper resistance.** The electromagnetic lock mechanism should be encased in hardened guard barriers to deter forced entry.

- **Mounting method.** The electromagnetic lock should be suitable for use with single and double doors with mortise or rim hardware and should be compatible with right- or left-hand mounting.

Bell or Alarm Box

This should be mounted on the front of the facility or a location where it will be in full view of neighbors and passersby. Such placement serves as a deterrent to many would-be burglars. The alarm should be placed high enough on the building to be out of easy reach.

Control Panels

Ideally, the control panels should be located close to the main entry and exit point. They should be positioned so that they cannot be reached without a ladder, should be close to a main electricity supply, and should not be attached to combustible material.

Passive Infrared (PIR) Detectors

Standard PIR detectors should not be mounted where they might be exposed to infrared light. Placement near windows, fires, filament lamps, and heat sources such as radiators and heaters could lead to nuisance alarms.

Door and Window Contacts

These are normally fitted to external doors and windows. However, they can be fitted to any vulnerable door or window to detect opening.

Shock Sensors

These are usually fitted to areas susceptible to forced entry, such as door or window frames. Door contacts detect the opening of a door or window but not necessarily breakage. If it seems possible that an intruder might attempt to gain access by kicking a panel out of a door or breaking a window, then shock sensors or PIR detectors may be a useful complement to door contacts.

Interconnection of Console Video Equipment

Between video equipment, the contractor should connect signal paths of 25 feet or less with RG-59/U coaxial cable; signal paths longer than 25 feet should use RG-11/U coaxial cable or fiber-optic cable. Cables should be as short as practicable for each signal path without causing strain at the connectors. Rack-mounted equipment on slide mounts should have cables of sufficient length to allow full extension of the slide rails from the rack. NFPA 731 provides more information on connecting equipment with "category" network cable.

Cameras

A camera needs a lens of the proper focal length to view the protected area. The contractor should do the following:

- Connect power and signal lines to the camera.

- If the camera has a fixed iris lens, set the camera to the proper f-stop to give full video level.

- Aim the camera to cover the alarm zone.

- For a fixed-mount camera installed outdoors and facing the rising or setting sun, aim the camera sufficiently below the horizon that the camera will not directly face the sun.

- Focus the lens to give a sharp picture over the entire field of view.

- Synchronize all cameras so the picture does not roll on the monitors when cameras are selected.

Chapter 7 of NFPA 731 provides details on selecting the appropriate location and lenses for cameras.

Exterior Fixed Mount

The contractor should install the camera mount as specified by the manufacturer and also do the following:

- Provide mounting hardware sized appropriately to secure the mount, camera, and housing with the maximum wind and ice loading encountered at the site.

- Provide a foundation for each camera pole as specified.

- Provide a ground rod for each camera pole, and connect the camera pole to the ground rod as specified.

- Provide electrical and signal transmission cabling to the mount location as specified.

- Connect signal lines and alternating current (AC) to mount interfaces.

- Connect a pole wiring harness to the camera.

Exterior Pan/Tilt Mount

The contractor should install pan/tilt mount, receiver/driver, and mount appurtenances as specified by the manufacturer and also do the following:

- Supply mounting hardware sized appropriately to secure the pan/tilt device, camera, and housing with the maximum wind and ice loading encountered at the site.

- Install pan/tilt control wiring as specified.

- Connect the pan/tilt device to control wiring and AC power.

Monitors

The contractor should install the monitors close to the operators' eye level or lower. The contractor should connect all signal inputs and outputs as recommended by the manufacturer, terminate video input signals, and connect the monitors to AC power.

Video Recording and Switching Equipment

The contractor should install the recording and switching equipment according to manufacturer's instructions and also do the following:

- Connect all subassemblies as specified by the manufacturer.

- Connect video signal inputs and outputs.

- Terminate video inputs as required.

- Connect alarm signal inputs and outputs.

- Connect control signal inputs and outputs for ancillary equipment or secondary control or monitoring sites as specified by the manufacturer.

- Load all software as specified and required for an operational CCTV system configured for the site requirements, including databases, operational parameters, and system, command, and application programs.

- Program the video annotation for each camera.

Other Features and Considerations

Conduit

All interior wiring—including low-voltage wiring outside the security center control monitoring console and equipment racks, cabinets, boxes, and similar enclosures—should be installed in rigid, galvanized steel conduit conforming to UL standards. Interconnection wiring between components mounted in the same rack or cabinet does not need to be installed in conduits. Minimum conduit size should be ½ inch. Connections should be tight-tapered and threaded. No threadless fittings or couplings should be used. Conduit enclosures should be cast metal or malleable iron with threaded hubs or bodies. Electric metallic tubing (EMT), armored cable, nonmetallic sheathed cables, and flexible conduit should normally not be permitted except where specifically required and approved by the customer. Data transmission media should not be pulled into conduits or placed in raceways, compartments, outlet boxes, junction boxes, or similar fittings with other building wiring. Flexible cords or cord connections should not be used to supply power to any components of the PPS except where specifically required and approved by the customer.

Grounding

All grounding must be in accordance with NFPA 70, articles 250 and 800. Additional grounding must meet manufacturers' requirements. All other circuits must test free of grounds. Grounding should be installed as necessary to keep ground loops, noise, and surges from adversely affecting system operation.

Enclosure Penetrations

All enclosure penetrations should be from the bottom unless the system design requires penetrations from other directions. Penetrations of interior enclosures involving transitions of conduit from interior to exterior, and all penetrations on exterior enclosures, should be sealed with an approved sealant to preclude the entry of water. The conduit riser should terminate in a hot-dipped galvanized metal cable terminator. The terminator should be filled with a sealant recommended by the cable manufacturer and in such a manner that the cable is not damaged.

Cold Galvanizing

All field welds and brazing on factory galvanized boxes, enclosures, and conduits should be coated with a cold galvanized paint containing at least 95 percent zinc by weight.

System Startup

The contractor should not apply power to the physical protection system until the following items have been completed:

- All PPS items have been set up in accordance with manufacturers' instructions.

- A visual inspection of the PPS system has been conducted to ensure that no defective equipment has been installed and that no connections are loose.

- System wiring has been tested and verified to be connected correctly.

- All system grounding and transient protection systems have been verified as properly installed and connected.

- Power supplies to be connected to the PPS have been verified as to the voltage, phasing, and frequency.

Configuration Data

The contractor should enter all data needed to make the system operational into the PPS database. The contractor should deliver the data to the customer on suitable forms. The data should include the contractor's field surveys and other pertinent information. The completed forms should be delivered to the customer for review and approval at least 30 days before database testing.

Graphics

Where graphics are required and are to be delivered with the system, the contractor should create and install the graphics needed to make the system operational. The contractor should use data from the contract documents, field surveys, and other pertinent information to complete the graphics. Graphics should have a sufficient level of detail for the system operator to assess the alarm. The contractor should also supply hard copy, color examples (at least 8 x 10 inches in size) of each type of graphic to be used for the completed system. The examples should be delivered to the customer for review and approval at least 30 days before acceptance tests.

Signal and Data Transmission System (DTS) Line Supervision

All signal and DTS lines should be supervised by the system. The system should supervise the signal lines by monitoring the circuit for changes or disturbances in the signal and for conditions described in UL 1076 for line security equipment. The system should initiate an alarm in response to a current change of 10 percent or greater. The system should also initiate an alarm in response to opening, closing, shorting, or grounding of the signal and DTS lines.

Housing

Sensors and system electronics need different types of housing depending on their placement:

- **Interior sensors.** Sensors to be used in an interior environment should be housed in an enclosure that provides protection against dust, falling dirt, and dripping non-corrosive liquids.

- **Exterior sensors.** Sensors to be used in an exterior environment should be housed in an enclosure that provides protection against windblown dust, rain and splashing water, hose-directed water, and ice formation.

- **Interior system electronics.** System electronics to be used in an interior environment should be housed in enclosures that meet the requirements of NEMA 250 Type 12.

- **Exterior system electronics.** System electronics to be used in an exterior environment should be housed in enclosures that meet the requirements of NEMA 250 Type 4X.

- **Corrosive settings.** System electronics to be used in a corrosive environment as defined in NEMA 250 should be housed in metallic enclosures that meet the requirements of NEMA 250 Type 4X.

- **Hazardous environments.** System electronics to be used in a hazardous environment should be housed in enclosures that meet the manufacturers' requirements for specific hazardous environments.

Nameplates

Laminated plastic nameplates should be provided for all major components of the system. Each nameplate should identify the device and its location within the system. Laminated plastic should be 1/8 inch thick and white with black center core. Nameplates should be a minimum of 1 x 3 inches, with minimum 1/4 inch-high engraved block lettering. Nameplates should be attached to the inside of the enclosure housing the major component. All major components should also have the manufacturer's name, address, type or style, model or serial number, and catalog number on a corrosion-resistant plate secured to the equipment. Nameplates are not required for devices smaller than 1 x 3 inches.

Tamper Switches

Enclosures, cabinets, housings, boxes, and fittings that have hinged doors or removable covers and that contain system circuits or connections and power supplies should be provided with cover operated, corrosion-resistant tamper switches, arranged to initiate an alarm signal when the door or cover is moved. The enclosure and the tamper switch should function together and should not allow a direct line of sight to any internal components before the switch activates. Tamper switches should do the following:

- Be inaccessible until the switch is activated.

- Have mounting hardware concealed so the location of the switch cannot be observed from the exterior of the enclosure.

- Be connected to circuits that are under electrical supervision at all times, irrespective of the protection mode in which the circuit is operating.

- Be spring-loaded and held in the closed position by the door or cover.

- Be wired so they break the circuit when the door or cover is disturbed.

Locks

For maintenance purposes, locks should be provided on system enclosures. Locks should be a UL-listed, round-key type with three dual, one mushroom, or three plain-pin tumblers, or a conventional key lock with a five-cylinder pin and five-point, three-position side bar. Keys should be stamped "DO NOT DUPLICATE." The locks should be arranged so that keys can only be withdrawn in the locked position. Maintenance locks should be keyed alike, and only two keys should be furnished. The keys should be controlled in accordance with a key control plan.

Wire and Cable

The contractor should provide all wire and cable not indicated as customer-furnished equipment. Wiring should meet NFPA 70 standards. The contractor should install the system in accordance with the standards for safety, NFPA 70, UL 681, UL 1037, and UL 1076 and the appropriate installation manual for each equipment type. Components within the system should be configured with appropriate service points to pinpoint system trouble in less than 20 minutes. The minimum conduit size should be ½ inch.

Local Area Network (LAN) Cabling

LAN cabling should be in accordance with the Electronic Industries Alliance (formerly Electronic Industries Association) standard EIA-568 A or B, category five.

Quality Assurance:

All work should conform to the following codes:

- Currently adopted National Electrical Code (NEC)

- Applicable federal, state, and local codes

- Currently adopted uniform building code

- Local electrical code as applicable

- Occupational Safety and Health Act (OSHA) standards

- Any additional codes effective at the job site

- Americans with Disabilities Act (ADA)

All materials should conform to the following codes:

- National Electrical Manufacturers Association (NEMA)

- American National Standards Institute (ANSI)

- Underwriters Laboratories, Inc. (UL)

Tuning the System

After installation, the system must be tuned to the operation of the facility. Otherwise, the system may generate too many unwanted alarms and confuse the operating personnel rather than assist them. Tuning the system requires knowing how the facility operates, what employees come and go, and what types of activities take place.

Time Periods for Alarms

To tune the system, the security manager should periodically run system reports and look at the alarm history, which shows nuisance alarms and alarm location, frequency, and timing. Patterns may emerge. For example, alarms may go off at certain times just because of day-to-day business. In those cases, the security manager can adjust the alarm operating times so that alarms will not be generated and security staff will not be unnecessarily distracted.

Responsibility for Monitoring Alarms

If an alarm associated with a loading dock door is constantly being received by the security monitoring center during business hours, then responsibility for monitoring of this alarm point should be transferred to personnel in that area. If alarms from mechanical or utility rooms are being received because maintenance personnel require access, procedures should be established to notify the central monitoring center that work will be performed in a certain area for a specific time, allowing the security systems operator to temporarily ignore those particular alarm points. Secured doorways where material movement is controlled must have a procedure such as a phone call to the central station. A security guard may be needed to control access to those areas.

Authorized Personnel

If authorized personnel are trusted and allowed to enter areas any time, then alarms should be shunted so that an alarm will not be generated.

Nuisance Alarms

Many nuisance alarms are caused by employee mistakes, such as opening the wrong doors, holding doors open, or forgetting to disarm alarm subsystems. Other alarms may be caused by malfunctioning door hardware. Signage, such as "keep doors closed," may help, as may adjusting guards' patrol times so they are more likely to catch instances where employees go through doors and leave them open. Patrol officers should check all the doors and make sure they are closed. The security manager should also examine the maintenance program to ensure that doors are kept in good operation. Maintenance should include frequent door inspections and prompt replacement of faulty components.

Improper Application

Sometimes the security and fire alarm system components selected are wrong for their application. For example, standard motion detectors should not be placed in a harsh environment, and microwave sensors should not be used in a room that has a hallway outside. These devices should be changed to eliminate nuisance alarms.

Maintaining the Operating Procedures

It is important to periodically review the operating procedures. Whenever procedures are changed, they should be documented with a new revision number and date. Saving the old revisions makes it possible to ascertain what policies and procedures were in effect at certain times—useful information if the security manager ever has to defend past actions. It is also important to align rewards and consequences. In other words, the security manager should reward people who do go a good job in security and make clear that there are consequences if operating procedures are not followed.

Incident response policies should be reviewed periodically by legal counsel. The legal review should ensure that procedures meet the following criteria:

- Are legally defensible and enforceable

- Comply with overall company policies and procedures

- Reflect known industry best practices demonstrating the exercise of due care

- Conform to national, state, and local laws and regulations

- Protect staff from lawsuits

In addition, legal counsel should consider the following:

- When to prosecute and what should be done to prosecute a person caught violating facility access rules

- What procedures will ensure the admissibility of evidence

- When to report an incident to local, state, or national law enforcement agencies

Legal counsel can help the security manager develop procedures and train security officers in such a way as to avoid problems that may lead to lawsuits over the following issues:

- **Failure to adhere to duty guidelines.** This occurs when officers engage in conduct beyond their established duties.

- **Breach of duty.** This occurs when officers engage in unreasonable conduct.

- **Proximate cause.** This term means an officer was the immediate cause of injury to a victim.

- **Foreseeability.** This term refers to events, especially those that could cause loss, harm, or damage, that the officers or management could have determined were likely to happen.

Failure to properly consider the human element and staff procedures when designing and installing new integrated security systems can turn a well-founded investment into an operational nightmare. Security managers can avoid this mistake by ensuring that their plans for security systems contain a complete analysis of how the systems will be operated and how the security force will respond to security incidents.

References

Angell, Grant, and Michael Baker (2002). *Limited energy reference manual*. Gladstone, OR: Limited Energy Resource Center.

NFPA Publications

National Fire Protection Association, 1 Batterymarch Park, Quincy, MA 02169-9101.

NFPA 70, *National electrical code*, 2002.

NFPA 730, *Guide for electronic premises security*, 2003 draft.

NFPA 731, *Standard for the installation of electronic premises security systems*, 2005.

NFPA 101, *Life safety code*, 2003.

NFPA 110, *Standard for emergency and standby power systems*, 2002.

NFPA 111, *Standard on stored electrical energy emergency and standby power systems*, 2001.

NFPA 5000, *Building construction and safety code*, 2003.

ANSI Publications

American National Standards Institute (ANSI), 25 West 43rd Street, 4th Floor, New York, NY 10036.

ANSI S1.4-1983 (R 2001) with Amd.S1.4A-1995, *Specification for sound level meters*.

SIA Publications

Security Industry Association, 635 Slaters Lane, Suite 110, Alexandria, VA 22314.

ANSI/SIA PIR-01-2000, *Passive infrared motion detector standard—features for enhancing false alarm immunity*.

ANSI/SIA CP-01-2000, *Control panel standard—features for false alarm reduction*.

UL Publications

Underwriters Laboratories Inc., 333 Pfingsten Road, Northbrook, IL 60062.

UL 294, *Standard for access control system units*, 1/29/1999.

UL 365, *Standard for police station connected burglar alarm units and systems*, 7/31/1997.

UL 606, *Standard for linings and screens for use with burglar-alarm systems*, 11/19/1999.

UL 608, *Standard for burglary resistant vault doors and modular panels*, 3/23/1999.

UL 634, *Standard for connectors and switches for use with burglar-alarm systems*, 2/23/2000.

UL 636, *Standard for holdup alarm units and systems*, 11/26/1996.

UL 639, *Standard for Safety for intrusion-detection units*, 2/21/1997.

UL 681, *Standard for installation and classification of burglar and holdup alarm systems*, 2/26/1999.

UL 1076, *Standard for proprietary burglar alarm units and systems*, 9/29/1995.

UL 1610, *Standard for central-station burglar-alarm units*, 10/26/1998.

UL 2017, *Standard for general-purpose signaling devices and systems*, 1/14/2000.

UL 2044, *Standard for commercial closed-circuit television equipment*, 6/26/1997.

UL 3044, *Standard for surveillance closed-circuit television equipment*, 12/19/1994.

U.S. Government Publications

Federal Communications Commission. Radio frequency devices. Title 47, *Code of federal regulations*, Part 15. Washington: U.S. Government Printing Office.

U.S. Army Corps of Engineers, Department of the Army (1991). *Closed-circuit television systems*. CEGS-6751.

U.S. Army Corps of Engineers, Department of the Army (1998). *Electronic security systems*. CEGS-3720.

CHAPTER 9:

TRAINING

While terrorist threats and natural disasters are deemed newsworthy, the activities that protect facilities are often considered mundane. Nevertheless, all the technological and procedural precautions in the world will be ineffective if they are not executed properly. Through well-conceived, well-executed security training programs, personnel can be better prepared to prevent incidents from happening, respond properly to incidents that do arise, and contribute to recovery efforts more effectively. Without appropriate training, personnel are more likely to contribute to security risks accidentally.

General Training Requirements

The customer should require that the installation contractor or systems integrator submit a proposal to conduct training courses for designated personnel in the operation and maintenance of the PPS. The training should address all the systems being installed. For example, if a CCTV system is being installed along with other systems, the CCTV training should be concurrent with and part of the training for the other systems.

Training manuals and training aids should be provided for each trainee, and several additional copies should be provided for archiving at the project site. The training manuals should include an agenda, defined objectives for each lesson, and a detailed description of the subject matter for each lesson. The contractor should furnish audiovisual equipment and other training materials and supplies. When the contractor presents portions of the course by audiovisual material, copies of the audiovisual material should be delivered to the customer in the same media used during the training sessions. The contractor should also recommend the number of days of training and the number of hours for each day. Approval of the planned training content and schedule should be obtained from the customer at least 30 days before the training.

All personnel giving instruction should be certified by the equipment manufacturer for the applicable hardware and software. The trainers should have experience in conducting the training at other installations and should be approved by the customer.

Training Topics

System Administration

This training focuses on determining and implementing system operational parameters and making any necessary operational adjustments. The first training class should be scheduled so that it is completed about 30 days before factory acceptance testing (if conducted) or site acceptance testing. By completing this training, system administrators will learn to use all system functions, including ID badge design and production; cardholder setup and access level assignment; access door programming; alarm setup and implementation; data storage and retrieval through reports; and system database backups. If CCTV systems are included in the PPS, the administrators will learn the architecture and configuration of the CCTV system; CCTV hardware specifications; and fault diagnostics and correction. A second training class should be conducted one week before the start of acceptance testing, and the system administrators should participate in the acceptance tests and reliability testing.

System Monitoring

This training focuses on day-to-day system operation. Upon completion of training, operators will know how to use system monitoring functions as determined by system administrative staff, including monitoring alarm events; monitoring personnel access to the facility; assessing, responding to, and clearing alarms and messages; monitoring access door status; and running routine reports. The first training class should be scheduled so that it is completed about 30 days before site acceptance testing begins. Upon completion of this course, each operator, using appropriate documentation, should be able to perform elementary operations with guidance and describe the general hardware architecture and system functionality.

This training should include the following topics:

- General PPS hardware architecture

- Functional operation of the system

- Operator commands

- Database entry

- Report generation

- Alarm assessment

- Simple diagnostics

A second training class should be conducted about one week before the acceptance test, and the system operators should participate in the acceptance tests and reliability tests. The course should include instruction on the specific hardware configuration of the installed subsystems and should teach students how to operate the installed system. Upon completion of the course, each student should be able to start the system, operate it, recover the system after a failure, and describe the specific hardware architecture and operation of the system.

Alarm Assessment and Dispatch

This training teaches PPS operators to assess the cause of different alarm conditions and properly deal with them. Before this training is conducted, the customer and contractor should have developed the alarm assessment and response procedures discussed in Chapter 5. This training should be based on the alarm types that might be expected from the various PPS subsystems.

Incident Response

This training provides instruction to the security response force on responding to different alarms and scenarios. Before this training is conducted, the customer and the contractor should have developed the incident response procedures discussed in Chapter 5. This training should be based on the various scenarios that the response force might encounter when responding to an alarm condition.

System Troubleshooting and Maintenance

This training focuses on the internal workings of the PPS so that students can troubleshoot and repair most problems. Topics in this class include system networking communications and diagnostics; device configurations and programming; controller setup, wiring, and diagnostics; software troubleshooting; and device programming. The system maintenance course should be taught at the project site about two weeks prior to reliability testing, and these students should participate in the reliability tests. The training should cover the following:

- Physical layout of each piece of hardware

- Troubleshooting and diagnostic procedures

- Repair instructions

- Preventive maintenance procedures and schedules

- Calibration procedures

IT Functions

This training is for personnel in the IT department who need to understand how the security system functions within a LAN/WAN network infrastructure. Topics in this class include network topologies and communications specific to each security subsystem; the impact of system functions such as digital video storage on network bandwidth; and the maintenance of data security.

System Overview

This training shows how the system will help meet overall security goals and objectives; how the system has been customized to meet operational requirements; and how to communicate security awareness to all employees.

References

ASIS International (2002, 2003). *ASIS International glossary of security terms.* Retrieved from www.asisonline.org/library/glossary/index.xml.

ASIS International (2003). *Private security officer selection and training.* Retrieved from www.asisonline.org/guidelines/guidelinesprivate.pdf.

Chapter 10:

Testing and Warranty Issues

The tests performed by the implementation team may involve equipment, personnel, procedures, or any combination of these. The ideal acceptance tests stress the system up to the established limits of site-specific threats. Tests should simulate actual threat conditions and provide conclusive evidence about the effectiveness of the security system.

Equipment performance testing is designed to determine whether equipment is functional, has adequate sensitivity, and will meet its design and performance objectives. It is not sufficient for a component to meet the manufacturer's standards if the component proves ineffective during testing. Equipment performance tests must always be coordinated with appropriate facility personnel.

Personnel performance tests are intended to determine whether procedures are effective, whether personnel know and follow procedures, and whether personnel and equipment interact effectively.

Some personnel performance tests require that personnel be tested without their knowledge. Particular care must be exercised to ensure that these types of tests are well-coordinated and safety factors carefully considered.

This chapter describes four types of tests:

- Pre-delivery or factory acceptance tests

- Site acceptance tests

- Reliability or availability tests

- After-acceptance tests

In determining what tests to conduct on security systems, several factors should be considered:

- Prioritizing of site-specific threats

- Identification of worst-case scenarios (lowest probability of detection, shortest amount of delay, various pathways into a facility)

- Identification of system functions (detection, assessment, delay) that are most critical in protecting company assets

- Determination of each subsystem's assumed detection probabilities and vulnerability to defeat

- Determination of the time for assessment of incidents (immediate assessment versus delayed assessment)

- Identification of the last possible points at which an adversary must be detected to allow adequate response by the facility protective force

- Comparison of vulnerabilities against findings and resolution of past security inspections and incidents

Generally, original copies of all data produced during factory, site acceptance, and reliability testing should be turned over to the customer at the conclusion of each phase of testing, prior to approval of the test.

The customer should provide documentation to the equipment supplier or system integrator describing the testing that must be accomplished during the installation and commissioning of the system. This documentation describes the personnel, equipment, instrumentation, and supplies necessary to perform acceptance testing. This documentation also describes who will witness all performance verification and reliability testing. The contractor should be informed that written permission of the customer should be obtained before proceeding with the next phase of testing.

This chapter also discusses the related concept of warranty issues.

Factory Acceptance Testing

Depending on the size and complexity of the system, the customer may require the contractor to assemble a test system including some or all of the system components, and then conduct tests to demonstrate that system performance complies with specified requirements in accordance with approved factory test procedures. The tests may be designed by the customer, or the customer may require the contractor to design the tests. The tests should be scheduled in advance of any installation of the new system, and the customer should attend and observe the tests. Model numbers of components tested should be identical to those to be delivered to the site. Original copies of all data produced during factory testing, including results of each test procedure, should then be delivered to the customer at the conclusion of factory testing for approval of the test. The test report should be arranged so that all commands, stimuli, and responses are correlated to allow logical interpretation.

The factory test setup should include the following:

- All security control center monitoring equipment

- At least one of each type of data transmission link, along with associated equipment, to provide a representation of an integrated system

- A number of local processors (field panels) equal to the number required by the site design

- At least one sensor of each type used

- Enough sensor simulators to provide alarm signal inputs (generated manually or by software) to the system equal to the number of sensors required by the design

- At least one of each type of terminal device used

- At least one of each type of portal configuration with all facility interface devices as specified

Equipment for testing CCTV systems includes the following:

- At least four video cameras and each type of lens specified

- Three video monitors

- Video recorder (if required for the installed system)

- Video switcher, including video input modules, video output modules, and control and applications software (if required for the system)

- Alarm input panel (if required for the installed system)

- Pan/tilt mount and pan/tilt controller if the installed system includes cameras on pan/tilt mounts

- Any ancillary equipment associated with a camera circuit, such as equalizing amplifiers, video loss/presence detectors, terminators, ground loop correctors, surge protectors, or other in-line video devices

- Cabling for all components

The customer should require a written report for the factory test indicating all the tests performed and the results. All deficiencies noted in the pre-delivery testing should be resolved to the satisfaction of the customer before installation and acceptance testing.

Site Acceptance Testing

The customer should require the contractor to develop a plan to calibrate and test all components, verify data transmission system operation, install the system, place the system in service, and test the system. Before conducting the site testing, the contractor should provide a report to the customer describing results of functional tests, diagnostics, and calibrations, including written certification that the installed, complete system has been calibrated and tested and is ready to begin site acceptance testing. This report should be received at least two weeks before the start of site testing. The report should also include a copy of the approved site acceptance test procedures.

Using the site acceptance test procedures, the contractor should demonstrate that the completed system complies with all the contract requirements. All physical and functional requirements of the project should be demonstrated. Through performance testing, the contractor shows system reliability and operability at the specified throughput rates for each portal, as well as the Type I and Type II error rates specified for the completed system. The contractor should calculate nuisance and false alarm rates to ensure that the system yields rates within the specified maximums at the specified probability of detection for each subsystem.

The site acceptance test should be started after written approval has been received from the customer. The contractor should be instructed that the customer may terminate testing any time the system fails to perform as specified. Upon successful completion of the site acceptance test, the contractor should deliver test reports and other documentation to the customer before commencing further testing.

For the PPS acceptance tests, the following should be done:

- Verification that the data and video transmission system and any signal or control cabling have been installed, tested, and approved as specified

- When the system includes remote control/monitoring stations or remote switch panels, verification that the remote devices are functional, communicate with the security monitoring center, and perform all functions as specified

- Verification that the video switcher is fully functional and that the switcher software has been programmed as needed for the site configuration

- Verification that all system software functions work correctly

- Operation of all electrical and mechanical controls and verification that the controls perform the designed functions

- Verification that all video sources and video outputs provide a full bandwidth signal

- Verification that all input signals are terminated properly

- Verification that all cameras are aimed and focused properly

- Verification that cameras facing the rising or setting sun are aimed sufficiently below the horizon that they do not view the sun directly

- If vehicles are used near the assessment areas, verification of night assessment capabilities (including whether headlights cause blooming or picture degradation)

- Verification that all cameras are synchronized and that the picture does not roll when cameras are switched

- Verification that the alarm interface to the intrusion detection subsystem is functional and that automatic camera call-up is functional for all designated alarm points and cameras

- When pan/tilt mounts are used in the system, verification that the limit stops have been set correctly, that all controls for pan/tilt or zoom mechanisms are operative, and that the controls perform the desired function

- If pre-position controls are used, verification that all home positions have been set correctly and have been tested for auto home function and correct home position

The contractor should deliver a report describing results of functional tests, diagnostics, and calibrations, including written certification that the installed, complete system has been calibrated and tested and is ready for reliability testing. The report should also include a copy of the approved acceptance test procedures.

Reliability or Availability Testing

Reliability testing is best conducted in alternating phases of testing and evaluation to allow for validation of the tests and corrective actions. The reliability test should not be started until the customer notifies the contractor, in writing, that the acceptance testing has been satisfactorily completed, training (if specified) has been completed, and all outstanding deficiencies have been corrected. The contractor should provide one representative to be available 24 hours per day, including weekends and holidays (if necessary), during reliability testing. The customer should terminate testing whenever the system fails to perform as specified.

Phase I Testing

The reliability test should be conducted 24 hours per day for 15 consecutive calendar days, including holidays, and the system should operate as specified. The contractor should make no repairs during this phase of testing unless authorized by the customer in writing. If the system experiences no failures during Phase I testing, the contractor may proceed directly to Phase II testing after receipt of written permission from the customer.

Phase I Assessment

After the Phase I testing, the contractor should identify all failures, determine causes of all failures, repair all failures, and deliver a written report to the customer. The report should explain in detail the nature of each failure, corrective action taken, and the results of tests performed; it should also recommend when to resume testing. About a week after receiving the report, the customer should convene a test review meeting at the job site to discuss the results and recommendations. At the meeting, the contractor should demonstrate that all failures have been corrected by performing appropriate portions of the acceptance tests. Based on the contractor's report and the test review meeting, the customer may set a restart date or may require that Phase I be repeated. If the retest is completed without any failures, the contractor may proceed directly to Phase II testing after receiving written permission from the customer. Otherwise, the testing and assessment cycles continue until the testing is satisfactorily completed.

Phase II Testing

Phase II testing should be conducted 24 hours per day for 15 con-secutive calendar days, including holidays, and the system should operate as specified. The contractor should make no repairs during this phase of testing unless authorized by the customer in writing.

Phase II Assessment

After the conclusion of Phase II testing, the contractor should identify all failures, determine causes of failures, repair failures, and deliver a written report to the customer. The report should explain in detail the nature of each failure, corrective action taken, and results of tests per-formed; it should also recommend when to resume testing. About a week after receiving the report, the customer should convene a test review meeting at the job site to discuss the results and recommenda-tions. At the meeting, the contractor should demonstrate that all fail-ures have been corrected by repeating any appropriate portions of the site acceptance test. Based on the contractor's report and the test re-view meeting, the customer may set a restart date or may require that Phase II testing be repeated. The contractor should not commence any required retesting before receiving written notification from the cus-tomer. After the conclusion of any retesting, the Phase II assessment should be repeated.

After-Implementation Testing

Several tests can be conducted after implementation, such as these:

- **Operational tests.** Operational tests are performed periodically to prove correct system operation but do not involve verification of equipment operating specifications, such as detection pat-terns of motion sensors or the exact distance a protected door is opened before alarming. Operational tests might check whether alarms activate correctly when protected doors are opened, whether motion sensors are activated when people walk in par-ticular locations, or whether tamper switches or duress buttons work properly.

- **Performance tests.** Performance tests verify that equipment conforms with equipment or system specifications. These tests determine parameters such as probability of detection and may require measuring devices, calibrated instruments, or special testing methods.

- **Post-maintenance tests.** Post-maintenance tests are operational tests conducted after preventive or remedial maintenance has been performed on security systems to make sure the systems are working properly and according to specifications.

- **Subsystem tests.** Subsystem tests ensure that large parts of the system are all working together as originally designed. Coordinated portions might include detection with normal response and detection with delays.

- **Limited scope tests.** Limited scope tests are used to test a complex system, which is broken down into several subsystems or segments that are tested separately. This type of testing is useful when it is difficult and time-consuming to test the entire system at one time.

- **Evaluation tests.** Evaluation tests are periodic, independent tests of the PPS to validate the vulnerability analysis and ensure that overall effectiveness is being maintained. An evaluation test should be performed at least once a year.

Warranty Issues

The contractor should be required to repair, correct, or replace any defect for a period of 12 months from the date of issue of the certificate of practical completion. The common time for the contractor to report to the job site to address a warranty issue is within four hours of the problem report. Moreover, the contractor should hold a sufficient stock of spares to allow speedy repair or replacement of equipment. Waiting for manufacturers to replace or repair equipment is not acceptable.

The contractor should provide the customer with telephone and fax contact numbers for reporting all problems and defects. The warranty should include full maintenance of equipment in accordance with the manufacturer's recommendations. The contractor should record all service visits in a database and provide report forms to the customer. The report form should record the date and time the fault was reported, the nature of the reported fault, the date and time of the visit, the actual fault identified, and the remedial work carried out.

A few questions to consider about warranties are as follows:

- Will the PPS supplier provide the warranty service, or will a third party do so?

- Are the service levels of the warranty service consistent with the system maintenance service levels?

- If items under warranty fail, what will happen with respect to the maintenance services any other parties are providing?

References

Garcia, Mary Lynn (2001). *The design and evaluation of physical protection systems.* (Pp. 277.) Boston: Butterworth-Heinemann.

Department of Energy (2002). *Physical security systems inspectors guide.* Washington.

CHAPTER 11:
MAINTENANCE, EVALUATION, AND REPLACEMENT

Organizations' increasing reliance on physical protection systems, coupled with the increasing scale and complexity of these systems, requires careful consideration of maintenance requirements. Software is never error-free, nor is hardware immune to electrical or mechanical failure. An organization's investment in security must therefore include maintenance services and a plan to minimize the potential for and impact of failures.

An effective maintenance program normally includes provisions that require facility technicians, augmented by contract representatives, to perform all tests, maintenance, calibrations, and repairs necessary to keep the physical protection systems operational. Frequent system failures, cursory testing procedures, and an inordinate number of components awaiting repair are all indications of a poor maintenance program. This chapter identifies the practical issues of hardware and software support and offers practical guidance for organizations considering a system maintenance agreement. Companies negotiating maintenance contracts should also seek legal advice as required. The chapter also raises the issue of evaluating whether and when to replace the physical protection system.

Physical protection system maintenance is of two main types:

- **Remedial maintenance.** This corrects faults and returns the system to operation in the event a hardware or software component fails. Remedial maintenance includes these measures:

    — Establishing a maintenance function that acts on and logs requests from users in the event of a system problem

    — Investigating the problem

    — Resolving the problem directly or managing the resolution if third-party service is required

    — Restoring the system or returning its use to the customer

    — Updating documentation with respect to the problem and its resolution

- **Preventive maintenance.** This consists of scheduled maintenance to keep the hardware and software in good operating condition. Preventive maintenance includes these activities:

    — Keeping electromechanical equipment (fans, filters, back-up batteries, door hardware, etc.) operating correctly

    — Replacing hardware components to keep the equipment up to current specifications (such as engineering changes)

    — Updating system and application software (bug fixes, new versions, etc.)

    — Testing and analyzing system reports (error logs, self-tests, system parameters, performance measures, etc.)

    — Maintaining system documentation

Normally, a system maintenance agreement includes both categories of services.

A PPS consists of hardware, software, networks, and services. It is the combination of all components working together correctly that provides the service to the users of the system. Failure of a single component may have no significant impact, or it may take the system down. A maintenance agreement should therefore be structured to resolve non-critical problems as well as issues that could cause major disruption to the organization and its business processes.

Common practice in the past has been for organizations to contract out the maintenance of their hardware, software, networks, and services separately. As systems have become more complex and integrated, the difficulties of identifying and resolving a problem or failure have increased. Not only is there the potential for finger-pointing between the parties over a problem, but the lost time of working through the various issues results in further frustration and delays.

Often the best solution is to select a single contractor to take responsibility for the maintenance requirements of the system. As the single point of contact, the contractor will diagnose the cause of the problem and manage the process of getting it resolved. Resolution may include third parties who supply or maintain particular system components, or it may require assistance from other service providers, such as telecommunication services or application software companies.

Remedial Maintenance

Maintenance Plan

Hardware and software system maintenance may be done by the equipment manufacturer, a system integrator, a maintenance contractor, the users, or any combination thereof. It is essential to develop guidelines to identify who is responsible for fault identification, problem diagnosis and verification, fault correction, repair testing, repair logging, and maintenance coordination and tracking. The coordination aspect is especially critical because security technologies may require several different types of maintenance skills depending on where a failure occurs.

It is also a good idea to train staff to perform preventive maintenance; this will help them better understand and operate the security systems. Such training is best provided by vendors as part of the procurement and installation phases of new systems. It is also useful to give technicians time to upgrade their skills and knowledge by exchanging information with fellow technicians during the installation. In addition, the maintenance plan should consider periodic "tuning" of the security system to each facility to eliminate nuisance and false alarms that create problems for the personnel monitoring and responding to the system.

When contracting for maintenance services, the customer and the contractor should do the following:

- Agree on the basis of the contract document.

- Document in detail the components of the systems that are to be maintained.

- Set out the service levels for each component or subsystem.

- Define roles and responsibilities of the parties to the agreement.

- Agree on pricing and payments.

- Set out how the agreement will be managed and administered.

Service Levels

The failure of various components will have varying levels of impact on the system. Failure of a single camera will have a smaller impact than failure of the communications server for the entire network. However, another workstation on the network may support an essential security service and require high-priority service.

The customer and the contractor will jointly need to develop a support plan and the appropriate service level and response times for each component. Components whose failure has a high impact on the system require a higher level of support. The extreme case would require that an engineer be stationed on-site with full spares at hand. It is more likely that the customer will require a guarantee of an immediate return phone call from a maintenance technician and a response to a site within two to four hours. The customer should consider and specify service levels that are realistic, measurable, and in accord with the organization's specific business needs, particularly if travel is involved. The costs for guaranteed response times of less than four hours can escalate rapidly due to the staff hours, travel, and equipment required.

On the other hand, there may be components of the system that the customer elects not to include under the full maintenance plan. Personal computer workstations may already be covered by a maintenance agreement with the computer supplier. However, excluding some items from maintenance or having other items on lower levels requires some careful thought .

Service levels and costs depend on the location of the system in relation to the supplier and on the ability to diagnose and fix problems remotely. Using a remedial maintenance provider based in another city may significantly extend response times. Requiring support outside normal business hours also affects service levels and costs.

Roles and Responsibilities

The major goal of system maintenance agreements is to ensure that the security system operates at its optimum capability with minimum downtime. Another goal is to minimize the number of different parties involved in managing the maintenance program. Roles and responsibilities of all of the parties providing services must be clearly defined, documented, and agreed upon with the system maintenance supplier.

In some cases, the supplier of the maintenance service is also the supplier of the hardware and software or an agent of that supplier. However, usually the systems integrator takes responsibility for ongoing maintenance as the prime contractor.

In establishing a system maintenance agreement, it is necessary to develop a plan that denotes the responsibilities of all parties, establishes the company's central point of contact, and facilitates agreements between the parties. The following parties may be involved:

- Hardware manufacturer or supplier

- Systems integrator

- Supplier of system tools and utilities where these are not provided by the hardware supplier

- Supplier of the application software

- Building owner (for such building services as power, water, and telephone/data circuits)

- Air conditioning service provider (in relation to specific equipment in the security monitoring room)

- Uninterruptible power supply (UPS) or emergency generator agent

- LAN and WAN equipment and service provider

- Telecommunications service provider (for phones, leased data circuits, and private/public network access)

- PC supplier

- Cabling supplier

The ideal may be for the prime vendor or systems integrator to manage all these parties in resolving faults or undertaking scheduled maintenance. Practically, that may not always be appropriate or cost-effective. Each customer may need to include or exclude specific third-party responsibilities.

Prices and Payments

Maintenance contractors usually have a scale of fees for the support of their products and for the delivery of their services. These fees may be arrived at from a complex mix of factors, including the complexity of the PPS, the cost of spare parts, the estimated number of failures per annum, the product usage frequency, the number of PPS users, and the age of the PPS. For high volume or standard systems, the fee may simply be a set percentage of the purchase price. Support fees may also be affected by the geographic location of the system or the ability for on-line diagnosis and support. Alternatively, it may be agreed that travel and accommodation costs will be billed separately.

Economies of scale may also affect a supplier's pricing for maintenance support. A larger number of units or customers in a geographic location may provide the opportunity to pass on savings in travel, spare parts inventories, staff, training costs, and establishment costs.

Similarly, the payment cycle for maintenance costs may vary according to the scope and nature of the service required. One approach might consist of a fixed fee for an advance period (month, quarter, or year) plus an allowance or a formula for the following:

- Discounts for the economies of longer-term contracts

- Credits when target response times are not met

- Additional costs associated with travel, accommodation, or work not covered by the agreement

- Call-outs outside agreed business hours

Over time, the factors that dictate maintenance pricing change. Some may decrease, but the majority increase along with inflationary pressures or the aging of the products. It is usual for pricing review milestones to be built into a maintenance agreement and for there to be an understanding between the parties as to the size of any increases at these milestones. Not typically covered in a maintenance agreement are such items as misuse, vandalism, lack of training due to turnover, acts of God, etc.

Administration

System maintenance takes place in an environment that changes over the term of the contract. The agreement itself needs to be monitored and maintained to reflect such changes. The security manager should regularly review the agreement, measure the provider's performance, and address the agreement's scope. The review should cover the following issues:

- Supplier performance against service levels and system performance for the previous period

- Call logging and account management

- Changes to the services or service levels that are required by the customer or recommended by the supplier

- Changes to the list of equipment or software on the system

- Customer's future plans for the system (including staffing, new developments, upgrades, special events, or changing priorities)

Documentation

The manufacturer or systems integrator should provide comprehensive documentation regarding the configuration of the system and all components, including switch settings, cable diagrams, spare parts lists, and installation steps. It is important that all subsystems have advanced levels of diagnostics that will identify faulty components so that they can easily be replaced in the field. For a large, decentralized system, the ability to conduct remote diagnostics is especially helpful. Subscribing to an upgrade service for the hardware and software after installation guarantees that the latest engineering change orders and field change orders will be incorporated into the system, thereby extending the system's life.

Records

Keeping accurate records about the security systems—especially maintenance and operator records—can help the security manager in many ways. Knowing what parts are failing or causing operator problems can help identify trouble spots and deficiencies. Keeping track of costs helps justify replacing unreliable systems.

Maintenance Records

Maintenance records of all components, cross-referenced to subsystems, should be kept to identify repair patterns. These records may point to components that should be closely inspected during preventive maintenance. The maintenance contractor (or whoever does the system maintenance) should keep records and logs of each maintenance task and should organize cumulative records for each major component and for the complete system chronologically. A continuous log should be maintained for all devices. The log should contain calibration, repair, and programming data. Complete logs should be kept and made available for inspection on-site, demonstrating that planned and systematic adjustments and repairs have been accomplished for the security system.

System Operator Records

System operator records should be maintained to identify problems that operators have with certain subsystems or components. These reports should be analyzed periodically to identify problem subsystems and components and to update operating procedures.

Spare Parts

It is useful to procure spare parts and repair equipment in advance (perhaps as part of the original device procurement) to minimize downtime in the event remedial repairs are required. The appropriate quantity of spares on hand varies according to the time required to obtain spares, the cost of maintaining inventory, and the likelihood of replacement. As a rule of thumb, about 5 percent of the capital cost of equipment for a location should be allocated each year for spare parts purchases. Spare parts inventories should reflect vendor recommendations. Standardization of devices, through sole-source vendor relationships or tight procurement specifications, can reduce inventory needs as well as training needs. A centralized budget is recommended for paying for unexpected replacement of devices.

Maintenance Manuals

The contractor should provide the customer with a manual that describes maintenance for all equipment, including inspection, periodic preventive maintenance, fault diagnosis, and repair or replacement of defective components.

Preventive Maintenance

Checklists should be developed to ensure that preventive maintenance tasks are performed adequately, and the checklists should incorporate any guidelines from equipment manufacturers. Preventive maintenance applies to most elements of the system infrastructure and includes such tasks as bulb replacement and camera lens cleaning.

Budgeting and resource allocation decisions must take into account not only security technicians but also information technology support. To conserve travel time, preventive maintenance activities should be pursued simultaneously with remedial maintenance activities to the extent possible. The following are typical tasks in preventive maintenance:

- Inspect the cabinets to ensure that voltage warning signs exist on equipment like power supplies.

- Ensure that security system warning signs, if installed, are in their proper location.

- Inspect enclosures for damage, unauthorized openings, and corrosion of metallic objects. Repair and paint as required.

- Inspect air passages and remove any blockage.

- Inspect, investigate, and solve conditions for unusual odors.

- Inspect locking devices. Repair as required.

- As equipment is operated and tested, listen to, investigate, and solve conditions for unusual noises.

- Inspect equipment mounting for proper installation.

- Inspect for loose wiring and components.

- Inspect electrical connections for discoloration or corrosion. Repair as required.

- Inspect electrical insulation for discoloration and degradation. Repair as required.

- Inspect equipment grounding components such as conductors and connections. Repair as required.

- Clean equipment. Remove debris, dirt, and other foreign deposits from all components and areas of non-encapsulated equipment, such as ventilated control panels.

- Tighten electrical connections.

- Torque all electrical connections to the proper design value.

- Perform operational tests periodically to prove correct subsystem operation, not necessarily to verify equipment operating specifications.

- Open protected doors.

- Walk into protected rooms.

- Test metal detectors by passing metal through the detection area.

- Prove operation of fence disturbance sensors by shaking the fence.

- Conduct visual checks and operational tests of the CCTV system, including switchers, peripheral equipment, interface panels, recording devices, monitors, video equipment electrical and mechanical controls, and picture quality from each camera.

- Check operation of duress buttons and tamper switches.

Adjustments

Periodic adjustments to security systems may have to be made to ensure that they are operating effectively. Detection patterns for motion sensors may have to be adjusted based on results of testing activities. Adjustments may need to be made to varifocal lenses on CCTV cameras to ensure that the proper scenes are being viewed.

Backup Equipment

Since security subsystems require power, an auxiliary power source consisting of batteries or generators must be available. Switchover must be immediate and automatic if the primary power source fails. In most cases, immediate and automatic switchover will not occur if a generator is the sole source of backup power; batteries are required, and the generator assumes the role once it obtains full power. To ensure effective operation of all devices, security managers should provide for a regular test and maintenance program. Such a program includes periodic testing of equipment and circuits including backup power, as well as thorough inspection of equipment and circuits by qualified service personnel. Records of these tests should include the test date, name of the person conducting the test, and results.

Evaluation and Replacement

At some point, the system will complete its useful life and the process of replacement will begin. To justify the replacement cost, the security manager should consider such factors such as the cost of maintenance, lack of spare parts, obsoleteness of hardware and software, operating costs, and unreliability. Replacement may also be justified by new technologies and features that provide improved security, the ability to reduce manpower, or other benefits.

The security manager should form a team of stakeholders in the organization, including members of the company's IT group, to select a system that will meet all stakeholders' needs. Performance deficiencies in the old system, such as the inability to read multiple card technologies or poor system response time, need to be addressed. Possible future uses of the system and ID cards, such as a debit card function in the employee cafeteria, should also be incorporated.

The team should build in considerable expansion potential to accommodate future plans for additional sites, panels, and cards. The team should also begin gathering information from reputable companies supported by a nationwide network of integrators. It is also crucial to make sure the system's software will pass muster with the IT department, which would have to work with it, and with the human resources department, which will need a seamless interface between its employee database software and the security system software.

References

Department of the Army (2001). *Maintenance of mechanical and electrical equipment at command, control communications, intelligence, surveillance, and reconnaissance (C4ISR) facilities.* Manual TM 5-692-1, Chapter 26.

INDEX